Preface

This series of books has been designed to cover the main aspects of physics courses that, in the U.K. at least, are generally taken by students aged sixteen to eighteen years, often prior to going to university. The series, while based upon the requirements of modern 'A'-level syllabuses, reflects the shift of emphasis in the teaching of physics in recent years and maintains a careful balance between the best of traditional courses and recent innovations. By selecting books from the series, and possibly supplementing with additional specialist material, a wide variety of courses can be covered. Some books, for example, could well be used in some of the technician courses in colleges.

Each book in the series is designed to cover a main topic in physics and each has been written in a feasible teaching sequence which carefully develops the structure of the physics. The chapters have been written in an essentially self-teaching method with text and questions interwoven. All the questions used within the text are supplied with suggestions for answers. At the end of each chapter are further questions, no answers supplied, which could be used for assessment. Many chapters also include background reading.

Though the physics in the books is developed from a basis of experimental data, details of experiments are not included. This is to enable teachers to use the book with the apparatus they have available and so plan the experimental work to suit their resources. References to sources of experiments are given.

The books have undoubtedly been influenced by my earlier work with the Nuffield Foundation Advanced Physics Teaching Project and UNESCO, as well as my present work with the Technician Education Council. The influence of the Physical Science Study Committee (PSSC) course in physics and the Project Physics Course is also apparent. The form of the books and the way the physics is presented is, however, my interpretation of the subject and any errors mine.

W. Bolton

Contents

Study Topics in Physics **Book 2**

Materials

W Bolton

Butterworths
London   Boston
Sydney   Wellington   Durban   Toronto

United Kingdom London	Butterworth & Co (Publishers) Ltd 88 Kingsway, WC2B 6AB
Australia Sydney	Butterworths Pty Ltd 586 Pacific Highway, Chatswood, NSW 2067 Also at Melbourne, Brisbane, Adelaide and Perth
Canada Toronto	Butterworth & Co (Canada) Ltd 2265 Midland Avenue, Scarborough, Ontario, M1P 4S1
New Zealand Wellington	Butterworths of New Zealand Ltd T & W Young Building, 77—85 Customhouse Quay, 1, CPO Box 472
South Africa Durban	Butterworth & Co (South Africa) (Pty) Ltd 152—154 Gale Street
USA Boston	Butterworth (Publishers) Inc 10 Tower Office Park, Woburn, Massachusetts 01801

First published 1980

© Butterworth & Co (Publishers) Ltd, 1980

ISBN 0 408 10653 0

British Library Cataloguing in Publication Data

Bolton, William
    Materials. — (Study topics in physics; Book 2).
    1. Materials
    I. Title II. Series
    620.1'1   TA403.2   79—41775

    ISBN 0—408—10653—0

Typset by Scribe Design, Gillingham, Kent
Printed and bound by Whitefriars Press Ltd, London & Tonbridge

1 Solids

Morris Marina 1300 L Coupé

Objectives

The intention of this chapter is to make you think about the wide
variety of materials that modern man uses and how their properties
affect the use. It is assumed that you can handle indices.

The general objectives for this chapter are that after working through
it you should be able to:

(a) State the definition of density and use it in problems;
(b) Recognise the terms tension and compression;
(c) State Hooke's law and use it in problems;
(d) Plot graphs of force/extension and from these graphs describe
the behaviour of different types of materials;
(e) Define stress and strain and calculate them from data;
(f) Plot graphs of stress/strain and from these graphs describe the
behaviour of different types of materials, using terms such as
elastic limit, Young's modulus, yield, tensile strength;
(g) Recognise that a relationship exists between Young's modulus
and the stiffness of a material;
(h) Use Young's modulus in calculations;
(i) Distinguish between brittle and ductile behaviour;
(j) Recognise the effect of stress concentrations on failure;
(k) Define the coefficient of linear expansion and use it in problems;
(l) Solve problems on the stresses produced when expansion or
contraction is restricted;
(m) Use the area and volume coefficients of expansion in problems.

Teaching note

Experiments appropriate to this chapter can be found in *Nuffield
Advanced Physics: Teachers' Guide 1*. Further reading can be found in
The New Science of Strong Materials and *Structures* both by J.E. Gordon,
and *The Use of Materials*, Engineering Science Project (Full bibliographic
details are given on p. 94.)

Which material?

The car shown opposite has a body made of thin sheets of steel. Why steel? Why not other materials? What are the properties required of a material that is to be used for the body of a car? How can we describe the behaviour of materials in order to know whether they have the required properties?

The photograph on the title page of this book shows a suspension bridge. The cables supporting the roadway are each made of 11 618 steel wires about 5 mm in diameter. What are the properties required of the material for it to be suitable for the cable of a suspension bridge?

You are probably sitting in a chair when you read this. What is the chair made of? What materials can chairs be made of? What are the properties required of the material for it to be suitable for a chair?

This chapter takes a look at some of the properties of solid materials and considers the properties in relation to how the materials will behave when used.

Question 1   *Figure 1.1* shows some tableware made from a plastic. What are the properties of the plastic that render it suitable for such a use? What other materials are used for tableware? What are the advantages and disadvantages of the different materials?

Figure 1.1 Makrolon® tableware

Density

A 1 kg block of aluminium and a 1 kg block of iron are quite different sizes even though the masses are the same. Every 1 kg block of aluminium does, however, have the same volume regardless of its shape. For a given material, the mass per unit volume is a constant. This constant is known as the **density**.

$$\text{Density} = \frac{\text{mass}}{\text{volume}}$$

Units: mass – kg, volume – m^3, density – kg/m^3 or kg m^{-3}.

The following are some typical values of densities.

Material	Density/kg m^{-3}
Steel	7800
Cast iron	7000
Aluminium	2700
Polythene	920
Cork	250

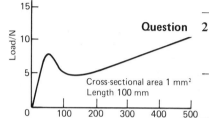

(a)

Question 2   What is the mass of a block of aluminium which has a volume of 1000 mm^3? What would be the length of the side of a cube of aluminium with this volume?

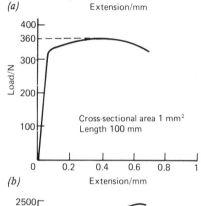

(b)

Elastic properties

Rubber bands stretch when pulled. A rubber eraser can be squashed a little. Stretching the rubber band results in an increase in the length of the rubber in the direction of the force and is said to be putting the material in **tension**. Squashing the rubber eraser causes its length to decrease in the direction of the force and is said to be putting the material into **compression**. It is not only rubber that can be put into tension or compression, all materials can.

How does the extension of a material depend on the force applied? *Figure 1.2* shows some experimental results. The force on the material was increased in gradual steps and a reading of the extension was taken at each step. This continued until the material broke.

The graphs show one common feature: an initial straight-line part where the extension is directly proportional to the load. In this region of the graph, doubling the load results in a doubling of the extension. This relationship of direct proportionality between extension and load is called **Hooke's law**. It can be represented by the equation

$$F = kx$$

where F is the applied force, x the extension and k a constant for the particular sample, often called the **force constant**.

In the case of the steel sample, *Figure 1.2(c)*, the extension is proportional to the force for loads up to about 2000 N. At this load the extension is about 0.1 mm. To produce half that extension, i.e., 0.05 mm, only about half the force, i.e., 1000 N, is needed. The force constant is about 20 000 N/mm, i.e., 2×10^7 N/m.

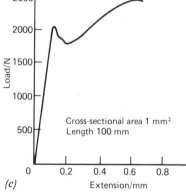

(c)

Figure 1.2 Load/extension graphs for samples of (a) polythene, (b) aluminium alloy, (c) steel

Question 3   Determine approximately from the graphs in *Figure 1.2(a)* and *(b)* the force constants for polythene and the aluminium alloy.

If you were to hold each end of the strip of polythene which gave *Figure 1.2(a)* and pull you would find that as you gradually increased your pull the sample would increase gradually in length until, at some particular force, it would suddenly increase considerably in length. This occurs at the point on the graph where the force is about 7.5 N and where the graph takes a downward turn. If, after this sudden increase in length, you continue to pull the polythene it continues to increase in length as the force is increased. You do, however, get a larger extension for a given increase in force than when you began to pull. The slope of the graph is more gradual than at the beginning. Eventually the polythene breaks.

If you had stopped pulling the polythene just after the sudden increase in length had occurred, or later, you would have found that the material did not return to its original length. If forces up to a certain value are applied to a material it will return to its original length when the applied force is removed; if greater forces are applied it does not. The material is said to be **elastic** when, on removal of the force, it returns to its original length. The departure from being elastic tends to occur at a point close to the limit of the initial straight-line part of the force/extension graph; for the polythene sample this occurs at about 6 N. If larger forces are applied there will be permanent deformation of the sample.

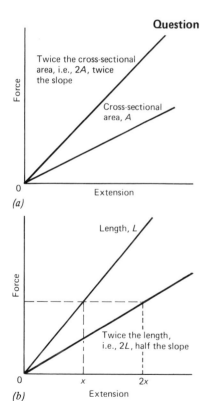

(a)

(b)

Figure 1.3 Force/extension graphs in the Hooke's law region for samples of (a) the same length but different cross-sectional areas, (b) the same cross-sectional area but different lengths

Question 4 Explain how the sample of aluminium alloy which gave *Figure 1.2(b)* would behave if you could pull the sample between your hands. The limit of elastic behaviour occurs at the limit of the straight-line part of the graph.

Force/extension graphs refer to samples of specific materials and with specific dimensions. The extension of a particular sample of material depends not only on the applied force but also on the cross-sectional area of the sample and its length. *Figure 1.3* shows results for a particular material.

For a given length of material, a sample with double the cross-sectional area needs twice the force to produce the same extension. If the cross-sectional area is trebled then treble the force is needed to give the same extension. For equal extensions the force per unit area is constant.

$$\text{Extension} \propto \frac{\text{force}}{\text{area}}$$

The force per unit area is called the **stress**. The area is that of the material before any forces are applied. (You may, however, come across a definition in which the area is that under load.)

$$\text{Extension} \propto \text{stress}$$

$$\text{Stress} = \frac{\text{force}}{\text{area}}$$

Units: force – N, area – m^2, stress – N/m^2 or pascal (Pa). $1 \text{ Pa} = 1 \text{ N/m}^2$.

For a given cross-sectional area doubling the length of the sample means that for a given force twice the extension is produced. With a constant force per unit area the extension per unit length is constant.

Extension ∝ original length

If the length is trebled then for the same force treble the extension is produced. The extension per unit original length is called the **strain**.

$$\text{Strain} = \frac{\text{extension}}{\text{original length}}$$

Units: extension – m, length – m, strain – no units.

For a given material, and irrespective of the dimensions of the sample, the same stress produces the same strain. Thus, a graph of stress against strain for a particular material is applicable to a sample of the material with any cross-sectional area or length. This enables us to determine the extension produced by a particular force from a knowledge of the cross-sectional area and length of the particular sample concerned.

Questions 5 A strip of polythene has a cross-sectional area of 1 mm² and is acted on by a tensile force of 5 N.
What is the stress acting on the strip?

6 A steel wire has a cross-sectional area of 1 mm² and is acted on by a tensile force of 1500 N.
What is the stress acting on the wire?

7 A strip of polythene 100 mm long extends by 50 mm under the action of forces.
What is the strain experienced by the sample?

8 A piece of steel wire 100 mm long extends by 0.05 mm under the action of forces.
What is the strain experienced by the wire?

Figure 1.4 shows the stress/strain graphs for a number of materials. These should be considered only as examples as the stress/strain graph for a particular material depends on the constitution of the material and the processes through which it has gone. In the graphs the stress is expressed as MN m⁻². This unit is meganewtons per square metre, a meganewton (MN) is a million newtons, i.e., 10^6 N. *Figure 1.5* shows the type of machine that is used to pull metal samples and the typical dimensions of a test piece for use in such a machine.

For the part of the stress/strain graph for which the strain is directly proportional to the stress, i.e., the region for which extension is proportional to force, the ratio stress/strain is given a special name – **Young's modulus of elasticity**.

$$\text{Young's modulus, } E = \frac{\text{stress}}{\text{strain}}$$

Units: stress – N/m², strain – no units, E – N/m², N m⁻² or pascal (Pa).

A high value of Young's modulus means that a large stress is needed to produce a given strain, i.e., a large force is needed for a particular size

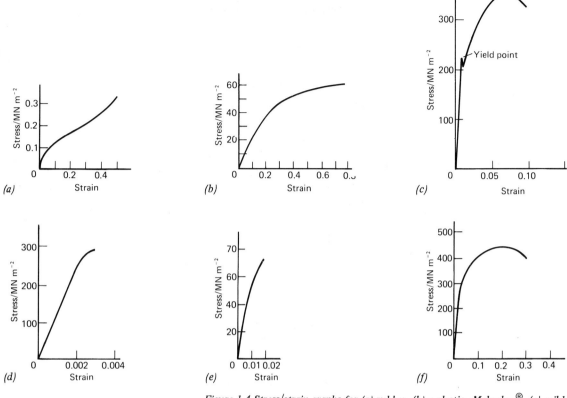

Figure 1.4 Stress/strain graphs for (a) rubber, (b) a plastic—Makrolon®, (c) mild steel, (d) cast iron, (e) beech wood, (f) brass

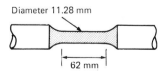

Diameter 11.28 mm

62 mm

Cross-sectional area 100 mm²

(a)

(b)

Figure 1.5 (a) Circular cross-section test piece used for tensile tests, (b) tensile testing machine

sample in order to produce an extension. The following are some typical values of this modulus.

Material	Young's modulus/GN m^{-2}
Mild steel	220
Cast iron	150
Brass	120
Beech wood (depending on grain direction)	10–20
Plastic, Makrolon®	2
Rubber	0.007

The above values indicate that, for the part of the stress/strain graph over which Hooke's law applies, the slope of the graph is steepest for the mild steel and least steep for the rubber. Look at the graphs in *Figure 1.4* and check this for yourself. (A giganewton (GN) is 10^9 N.)

Questions 9   A tensile test on a metal sample which had a cross-sectional area of 200 mm^2 and a length over which the extension was measured of 80 mm gave the following results. (A kilonewton (kN) is 10^3 N.)

Load/kN	0	10	20	30	40	50	60	70	80
Extension/mm	0	0.031	0.063	0.095	0.127	0.160	0.193	0.229	0.276

(a) Draw the stress/strain graph.
(b) Determine the value of Young's modulus.

10   A mild steel bar has a modulus of elasticity of 220 GN/m^2. If the cross-sectional area of the bar is 30 mm^2 calculate the force needed to produce an extension of 0.5 mm in a length of 100 mm.

11   A 0.50 kg mass is hung from the end of a vertical, suspended, wire of length 2.0 m and diameter 0.40 mm. If the material of the wire has a Young's modulus of 200 GN/m^2 what is the extension produced?

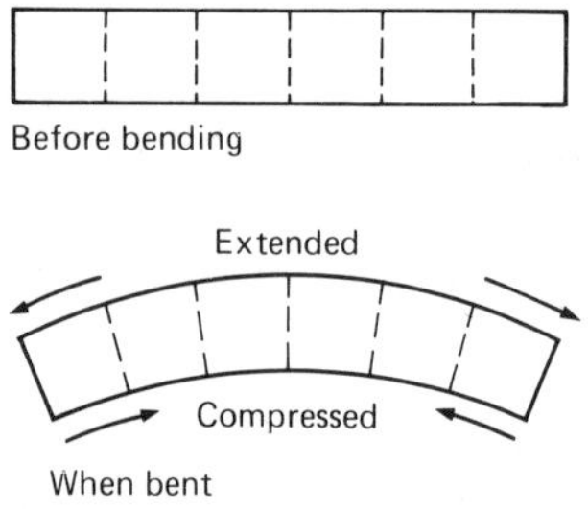

Before bending

Extended

Compressed

When bent

Figure 1.6 Bending a strip of material

If you take a strip of metal and apply forces to it in such a way as to bend it then one of the surfaces is made to extend, and so be in tension, and the other becomes compressed (*Figure 1.6*). For most materials the modulus of elasticity for tension is the same as for compression. The modulus of elasticity represents the ease with which the material can be extended, or compressed. A high value of the modulus means that a material is difficult to extend, or compress; a high force is needed. Bending a material stretches and compresses it; thus a material with a high modulus of elasticity is difficult to bend – it is stiff. The modulus of elasticity is a measure of the stiffness of a material.

Questions 12   Which would bend more easily – a strip of Beech wood or the same size strip of the plastic Makrolon®?

13 The modulus of elasticity of the plastic used for the tableware described in *Figure 1.1* is lower than that of the material used for the fired clay form of crockery.

How would you expect the behaviour of the two types of tableware to differ when the tableware is being used?

The term **limit of proportionality** is used for the stress at which the stress/strain graph ceases to be a straight line. The term **elastic limit** is used for the stress at which the strain ceases to be entirely elastic. If the strain is entirely elastic then the material should return exactly to its original size when the stress is removed. For most materials the elastic limit and the limit of proportionality have the same stress value. For some materials, e.g., many steels, the stress/strain graph has a sharp discontinuity (see *Figure 1.4(c)*). At this point the material suddenly yields with little or no increase in applied load. The **yield point** stress is close to or coincident with the elastic limit. If the stress applied to a sample of material is high enough the material breaks.

The maximum load applied divided by the original cross-sectional area of the sample is called the **tensile strength**. Thus, for the aluminium alloy which gave the load/extension graph in *Figure 1.2* the maximum load was 360 N and the original cross-sectional area 1 mm^2. This gives a tensile strength of 360 N/mm^2 or 360 MN/m^2. (1 N/mm^2 = 1 MN/m^2). The following are some typical tensile strengths.

Materials	*Tensile strength*/MN m^{-2}
Mild steel	400
Cast iron	300
Brass	400
Beech wood	60
Plastic, Makrolon®	60
Rubber	0.3

Questions

14 Calculate the maximum load that can be applied to a bar of mild steel with a cross-sectional area of 200 mm^2 before the bar breaks.

15 Determine the tensile strength for the polythene sample that gave the load/extension graph in *Figure 1.2(a)*.

If you break a china tea cup it is possible to pick up all the pieces and stick them back together again and still have something that looks like a cup. If you have a crash with a car and bash in the door, *Figure 1.7*, then you would not have a collection of pieces to stick together to make the original shape but you might take a hammer and beat the door out again to its original shape. The china tea cup material is said to be **brittle**, the mild steel of the car door **ductile**.

Figure 1.7 Mild steel dents. It is ductile

Brittle fracture of a material occurs when there is little, if any, permanent deformation of the material prior to fracture. Cast iron is a brittle material. The stress/strain graph for cast iron (*Figure 1.4(d)*) shows that the difference between the limit of proportionality, the elastic limit, and the final breaking point, is very small. No sooner does cast iron start to deform permanently than it breaks. In the case of the stress/strain graph for mild steel (*Figure 1.4(c)*) the difference between the elastic limit, the limit of proportionality, and the final breaking point is much greater and a considerable amount of permanent deformation of a mild steel sample can occur before it breaks.

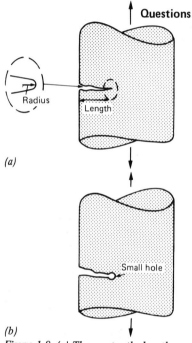

(a)

(b)

Figure 1.8 (a) The greater the length and the smaller the radius of a crack the greater the stress concentration, (b) drilling a hole at the end of the crack can reduce the stress concentration by increasing the radius

Questions

16 Is human bone a brittle or a ductile material?

17 The stress/strain graph for brass is given in *Figure 1.4(f)*. Would this suggest that brass is a brittle or a ductile material?

18 Is glass a brittle or a ductile material?

19 Is the material out of which paper clips are made a brittle or a ductile material? Try an experiment with a paper clip.

If you want to cut a sheet of glass the usual method is to cut the surface slightly and then either to give the sheet a sharp tap or to bend it slightly. The glass is not cut right through by the initial cut but this, relatively superficial, cut enables the sheet to be broken quite easily. A small notch in the surface of a material considerably affects the stress at which many materials break. The notch produces what are called **stress concentrations.** The deeper the notch and the smaller the radius of the tip the greater the stress concentration and thus the easier it becomes to break the material (*Figure 1.8*). One way of reducing the chance of a crack in a material leading to failure is to drill a small hole at the end of the crack and so increase the radius of the end of the crack.

Question

20 Why are sheets of postage stamps perforated? What do you think would be the optimum shape of the perforations from the point of view of easy separation?

The effect of temperature on solid materials

The mercury-in-glass thermometer is based on the idea that the volume of mercury increases when something we call temperature increases. The scale on the thermometer can be arrived at by fixing two points and then assuming that the expansion of the mercury is such that the change in volume per degree is constant and does not depend on the temperature, i.e., that there is a linear relationship between the volume of the mercury and temperature. Thus, if we take temperature to be that property defined by a mercury-in-glass thermometer then if we investigate how the expansion of a solid changes with temperature all we are doing is comparing the way the solid expands with the way the mercury in the glass expands.

In general, the amount by which a piece of material expands is found to be proportional to the temperature change causing the expansion. The expansion also depends on the initial dimensions of the substance. Thus, for the linear expansion of a material, we can represent the change in length by the equation

Change in length = coefficient of linear expansion × initial length
    × change in temperature

$$L_\theta - L_0 = \alpha L_0 \theta$$

where L_θ is the length after a temperature change of θ, L_0 is the initial length and α the **coefficient of linear expansion**. The value of the coefficient depends on the material concerned.

Units: $L - \mathrm{m}, \theta - {}^\circ\mathrm{C}, \alpha - {}^\circ\mathrm{C}^{-1}$.

Material	Coefficient of linear expansion/$^\circ C^{-1}$
Copper	16×10^{-6}
Steel	11×10^{-6}
Brass	19×10^{-6}
Glass, Pyrex	3×10^{-6}
Glass, soda	9×10^{-6}
Polythene	3×10^{-4}

Questions

21 Calculate the amount by which a bar of copper of length 500 mm might be expected to expand when the temperature of the bar rises by 80 $^\circ$C.

22 Which will expand more – a bar of steel or a bar of brass the same length – if both are heated through the same temperature change?

23 Which will expand more – a rod of Pyrex glass or a rod of soda glass the same length – if both rods are heated through the same temperature change?

24 Telegraph wires are fixed between two poles when the temperature is about 20 $^\circ$C.
If the wires are made taut between the poles what would you expect to happen when

(a) the temperature rises above 20 $^\circ$C and
(b) the temperature falls to below 20 $^\circ$C?

When thermal expansion or contraction is prevented, perhaps by the material being held between fixed end supports, considerable forces can be produced. *Figure 1.9* shows such a situation – a metal rod held between two fixed end plates. The taut telegraph wires between the poles in question **24** will exert strong forces on the poles when the temperature drops and the forces might be large enough to break the wires.

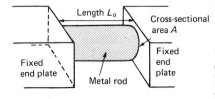

Figure 1.9

The initial length of the rod in *Figure 1.9* is taken as L_0. When the temperature rises by θ the rod would, if free, expand by an amount $L_0\theta\alpha$. This expansion is prevented. It is as if a rod of length $(L_0 + L_0\theta\alpha)$ is compressed to a length L_0. The compressive strain is thus

$$\text{Strain} = \frac{\text{change in length}}{\text{initial length}}$$

$$= \frac{L_0\theta\alpha}{L_0 + L_0\theta\alpha}$$

As the change in length $L_0\theta\alpha$ is small compared with L_0 the expression can be simplified to give

$$\text{Strain} = L_0\theta\alpha/L_0$$

$$= \theta\alpha$$

If the strain is such that the limit of proportionality between stress and strain is not exceeded, then

$$\frac{\text{Stress}}{\text{Strain}} = \text{Young's modulus}, E$$

and so

$$\text{Stress} = E\theta\alpha$$

The force exerted on the material is given by stress = force/area and so

$$\text{Force} = AE\theta\alpha$$

where A is the cross-sectional area of the rod.
The force exerted on the end plates is opposite and equal to this force.

Questions

25   A bar of steel of length 2 m is constrained between two rigid supports.
Calculate the stress developed in the bar as a result of a temperature rise of 10 °C.

26   What forces are required to stop a brass rod of length 200 mm and cross-sectional area 100 mm² expanding when the temperature rises by 5 °C?

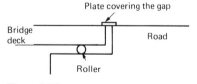

Figure 1.10

A bridge expands when the temperature rises and, if constrained, the temperature rise could result in buckling of the bridge girders. One end of a bridge is generally left free to move: if you look carefully you will be able to see the gap left to allow for expansion (*Figure 1.10*).

Area and volume expansion

When the temperature rises the area of a surface increases. The change in area can be represented by an equation similar to that for linear expansion

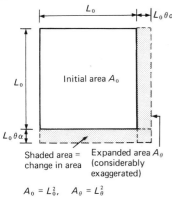

Shaded area = change in area

Expanded area A_θ (considerably exaggerated)

$A_0 = L_0^2$, $A_\theta = L_\theta^2$

Change in area = $L_0 \times L_0\theta\alpha$ $+ L_0 \times L_0\theta\alpha + (L_0\theta\alpha)^2$

Neglecting $(L_0\theta\alpha)^2$ as being insignificant

Change in area = $2L_0^2\theta\alpha$ $= A_0\theta(2\alpha)$

Figure 1.11 Derivation of coefficient of area expansion

Change in area = coefficient of area expansion × original area × change in temperature

$$A_\theta - A_0 = \beta A_0 \theta$$

where A_θ is the area after a temperature change of θ, A_0 is the original area and β a coefficient which depends on the material concerned.

Units: $A - \mathrm{m}^2, \theta - {}^\circ\mathrm{C}, \beta - {}^\circ\mathrm{C}^{-1}$

Values of the **coefficient of area expansion** are not usually quoted as each is twice the value of the corresponding linear coefficient of expansion. *Figure 1.11* shows the derivation of this relationship. Thus, for Pyrex glass the area coefficient is $2 \times 3 \times 10^{-6}\ {}^\circ\mathrm{C}^{-1}$, and for soda glass is $2 \times 9 \times 10^{-6}\ {}^\circ\mathrm{C}^{-1}$.

Question   **27**   Which area will expand more – a sheet of soda glass or a sheet of Pyrex glass with the same original area – if both are heated through the same temperature rise?

The volume of a solid increases when the temperature increases. The change in volume can be represented by an equation similar to that for linear or area expansion.

Change in volume = coefficient of volume expansion × original volume × change in temperature

$$V_\theta - V_0 = \gamma V_0 \theta$$

where V_θ is the volume after a temperature change of θ, V_0 is the original volume and γ the **coefficient of volume expansion** which depends on the material concerned.

Units: $V - \mathrm{m}^3, \theta - {}^\circ\mathrm{C}^{-1}, \gamma - {}^\circ\mathrm{C}^{-1}$

Values of the coefficients of volume expansion are not usually quoted as each is three times the value of the corresponding coefficient of linear expansion. *Figure 1.12* shows the derivation of this relationship. Thus for Pyrex glass the volume coefficient is $3 \times 3 \times 10^{-6}\ {}^\circ\mathrm{C}^{-1}$, and for soda glass is $3 \times 9 \times 10^{-6}\ {}^\circ\mathrm{C}^{-1}$.

$V_0 = L_0^3$, $V_\theta = L_\theta^3$

Change in volume = $3 L_0^2 \times L_0\theta\alpha + 3 L_0 \times (L_0\theta\alpha)^2$ $+ (L_0\theta\alpha)^3$

Neglecting $3 L_0 \times (L_0\theta\alpha)^2$ and $(L_0\theta\alpha)^3$ as being insignificant

Change in volume = $3 L_0^3 \theta\alpha$ $= V_0\theta(3\alpha)$

Figure 1.12 Derivation of coefficient of volume expansion

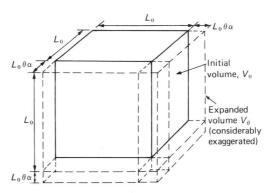

A hollow cube expands by just the same amount as a solid cube which has the same external dimensions, the length of each side of the cube expanding by the same amount regardless of whether it is solid or hollow. A hollow body such as a bottle or a beaker expands the same amount as a solid body having the same external measurements.

Questions

28   By how much will the volume of a soda glass bottle of volume 200 cm^3 expand when the temperature rises by 10 °C?

29   Which will expand more − a Pyrex or a soda glass bottle of the same volume − when both are subject to the same temperature change?

Suggestions for answers

1   A cup should be rigid and not deform when full of hot coffee or tea. It should be impenetrable to liquids. It should not taint the coffee or tea. You should be able to think of more requirements. Plastics, glass and fired clay are commonly used for tableware. Some materials break more easily than others, some deform more easily than others. The plastic used for the tableware in *Figure 1.1* is described by the manufacturer as of 'Outstanding strength, impact strength and hardness, able to withstand temperatures up to 145 °C and down to −150 °C, no health hazard, be dimensionally stable'. This question has no simple answer but requires of you an intelligent discussion of the problem.

2   $1000 \times 10^{-9} \times 2700 = 2.7 \times 10^{-3}$ kg = 2.7 g. A cube of side 10 mm.

3   Polythene 0.180 N/mm or 180 N/m; aluminium 5000 N/mm or 5×10^6 N/m.

4   As the force is gradually increased so the material gradually extends until suddenly quite a small increase in force produces a considerable extension. The material then extends very easily until it breaks. If, when the material starts to extend considerably, it is released, it will not return to its original dimensions but show a permanent extension.

5   5 N/mm^2 or 5 MN/m^2 (1 MN = 10^6 N)

6   1500 N/mm or 1500 MN/m^2 or 1.5 GN/m^2 (1 GN = 10^9 N)

7   50/100 = 0.5. Strain is sometimes written as a percentage and so this strain would be 50 %

8   0.05/100 = 0.0005 or 0.05 %

9   See *Figure 1.13*, about 120 GN/m^2

10   33 kN

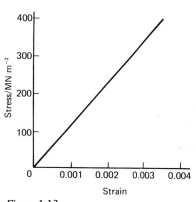

Figure 1.13

11 $F = 0.50 \times 10 = 5$ N; $A = \dfrac{\pi \times 0.0004^2}{4}$ m^2; 0.40 mm (to two significant figures)

12 The plastic, because it has a lower modulus

13 The plastic tableware would be more likely to bend when in use

14 80 kN

15 10 MN/m^2

16 Brittle, certainly when not that of a child. You can 'repair' broken bones by fixing them back together again.

17 Ductile

18 Brittle

19 Fairly ductile

20 So that the paper will tear along the right lines; so that the paper will 'break' easily. Long 'cracks' with pointed ends.

21 0.64 mm

22 Brass

23 Soda glass

24 The wires would (a) sag; (b) perhaps break under the forces involved

25 24.2 MN/m^2

26 1140 N

27 Soda glass

28 0.054 cm^3

29 The soda glass bottle

Further problems   No suggestions for answers are given for these problems.

30 When I take a strip of rubber in my hands and pull it, the strip extends fairly easily at first but when it has extended quite some way it becomes more difficult to stretch.
Sketch the force/extension graph for the rubber.

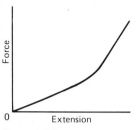

Figure 1.14

31   Describe how it would feel to stretch a strip of the material which gave the force/extension graph shown in *Figure 1.14*.

32   Hooke's law is often expressed as: the extension of a material is directly proportional to the force causing the extension.

(a) How would a graph of force against extension look for a material that obeyed Hooke's law?
(b) Write an equation linking force and extension for such a material.

33   When you drop a china cup and it breaks you can carefully glue all the pieces back together again and the cup is restored to its original shape. When cars collide and the metal wings are damaged the damage shows as deformation. The wings can be restored to their original shape by hammering.
How do the force/extension graphs for china and the metal differ?

34   (a) A rubber strip, cross-section 4 mm X 1 mm, is pulled by a force of 5 N. What is the stress acting on the rubber?
(b) A rubber strip, length 10 cm, is pulled and extends by 5 mm. What is the strain?

35   (a) If I take two bars, each of a different material but the same size, in my hands and flex them, how can I tell which material has the higher Young's modulus? Which material is the stiffer, the one with the higher Young's modulus or the one with the lower Young's modulus?
(b) Rubber has a low value of Young's modulus. What does this tell you about the behaviour of rubber?
(c) Young's modulus for spruce wood along the grain is 12 GN/m^2, and across the grain is 0.6 GN/m^2. If you flex a piece of spruce wood how could you tell, without looking at the grain direction, which was the direction with the higher Young's modulus?

36   The following data was obtained when a sample of a material was stretched.

Force/kN	0	10	20	30	40	50	60
Extension/mm	0	0.016	0.032	0.047	0.065	0.080	0.097

(a) Is Hooke's law obeyed for the results quoted?
(b) The extension was measured for a gauge length of 100 mm and the sample had a cross-sectional area of 300 mm^2.
What is Young's modulus for the material?

37   Two coat hangers were compared under load. Coat hanger A started to show a permanent bend when the load reached 300 N and finally became so deformed when the load was 450 N that the test had to be discontinued. Coat hanger B showed no permanent deformation until the load reached 400 N and then snapped when the load reached 410 N.

(a) Which coat hanger was made of a ductile material, and which of a brittle material?

(b) Make sketches of how you envisage the stress/strain graphs for the materials used for the two coat hangers.

38   A brass bar has a modulus of elasticity of 120 GN/m^2.
If the cross-sectional area of the bar is 200 mm^2, calculate the extension produced in a length of 300 mm when a force of 20 kN is applied.

39   A student carried out a stretching experiment on a copper wire of diameter 0.091 cm, measuring the extension over a length of 4.0 m. To keep the wire taut a load of 2 kg was suspended from the wire and the zero extension reading taken with this load. The following are the results.

Added load/kg   0 1   2   3   4   5   6   7
Extension/mm   0 0.5 1.0 1.5 2.0 2.5 3.0 7.5

(a) What was Young's modulus for the wire?
(b) What was the stress at the limit of proportionality?

40   What is the greatest load that can be suspended from a steel wire which has a cross-sectional area of 10^{-6} m^2 if it is not to break? The tensile strength of the steel is 400 MN/m^2.
How many of these wires would be needed to make a hawser capable of lifting masses of the order of 10 000 kg?

41   Human skin stretches very easily, under very low stress and with very little change in stress, up to strains of the order of 0.2. Beyond this strain much larger stresses are needed and beyond strains of about 0.4 very large stresses are needed to produce higher strains. Sketch the stress/strain graph for human skin.

42   Aluminium has a coefficient of linear expansion of 24×10^{-6} °C^{-1}.
By how much would a 1 m length of aluminium expand when the temperature changes by 10 °C?

43   The length of a bridge is 800 m. If it were a continuous steel span, free to move at each end, what would be the range of motion that would have to be allowed for between a cold winter day of −10 °C and a hot summer day of 30 °C? Take the coefficient of linear expansion to be 11×10^{-6} °C^{-1}.

44   What would be the stress produced in the bridge in question 44 on the cold winter day if the bridge has been rigidly fixed in place on the hot summer day? Take Young's modulus to be 200×10^9 N/m^2.

45   A 'bimetallic strip' consists of strips of copper and steel firmly fixed together along their lengths. What happens when the strip gets warmer? The two materials have different coefficients of expansion.

46   Why do glasses sometimes crack if hot water is poured into them? Glass is a poor conductor of heat.

Appendix 1.1   My nylon socks

My socks are made of nylon: you may find nothing strange or novel in that fact. When, however, I look at the labels on my clothes I find that almost all of them are made of artificial fibres. Yet it was not that many years ago when most clothes were made of cotton or wool. Nylon was only produced on a commercial scale in 1940, terylene in 1941. Now there are many different forms of artificial fibres.

My nylon socks do not need darning whereas this was a routine task when I wore woollen socks. I do not know what eventually happens to nylon socks — perhaps somewhere in my home there is a cupboard full of old nylon socks. I must confess that I do occasionally have to buy new socks but it is generally because the old ones seem to have disappeared.

Why do nylon socks last longer than woollen socks? Some of the physical properties of nylon in relation to the properties of other fibres are shown in *Figure 1.15*. Nylon has a higher tensile strength than the other fibres and holds its shape better. Nylon is stronger and more elastic. We begin to have some clue as to why my nylon socks last longer than woollen ones.

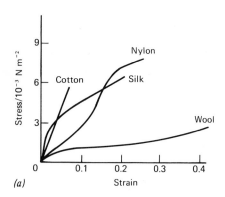

(a)

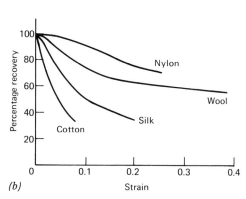

(b)

Figure 1.15 (a) Typical stress/strain graphs, (b) a 100% recovery would mean that the material was completely restored to its original length when the stress was released

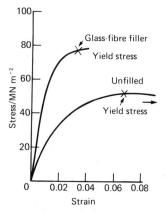

Figure 1.16 Stress/strain graphs for Makrolon® (a polycarbonate polymer) showing the effects of adding a filler

Appendix 1.2   Papier-mâché

An activity you might well have taken part in at some time in your life — perhaps in primary school — is to make shapes with papier-mâché. The first step in this technique consists of coating the object to be used as the mould with some substance that prevents paper sticking to it, possibly soap or oil. Pieces of paper, perhaps newspaper, are then torn into strips and soaked in glue or a paste, and the sodden strips pressed on to the object. This is continued until the required thickness has been built up. After thorough drying the paper shell can be removed from the mould, and then painted. The resulting material is surprisingly strong — certainly much stronger than the paper by itself, or the solid glue.

The technique of making papier-mâché shapes is not so different from that used for making fibre-glass boats. Glass-fibre fabric is bonded

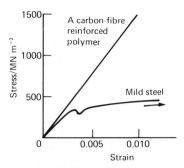

Figure 1.17 *Comparison between a typical carbon-fibre reinforced polymer and mild steel*

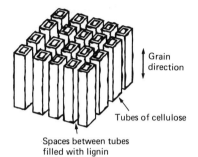

Figure 1.18 *The composite known as wood*

in layers on to a mould by means of cold-setting resin adhesives. Such a method can be used to produce quite complex shapes.

The papier-mâché and the fibre-glass materials can both be thought of as composite materials. Composites are combinations of materials in which each material retains its individual identity. The papier-mâché can be considered as glue containing paper as a 'filler'. The fibre-glass boat can be considered as the resin, a polymer, containing glass fibres as 'filler'. The incorporation of the 'filler' considerably increases both the strength and the stiffness. It also produces a material which is tough. A tough material is one that resists crack propagation. A crack propagating through the composite is blocked and blunted by encountering the filler.

The mechanical properties of polymers are often not ideal for many applications — some polymers are not stiff enough, others are too brittle, almost all are not tough enough. The use of fillers enables new materials to be produced which have much better mechanical properties. *Figure 1.16* shows the effect on the stress/strain graph of using glass-fibre reinforcement in a polymer (polycarbonate). The glass fibre increases the Young's modulus and the yield point. Carbon fibres used as fillers for polymers produce very strong and stiff materials — stronger and stiffer than steel (*Figure 1.17*).

Composites are not a modern invention, the Ancient Egyptians made the mummy cases out of papyrus moulded to a shell rather like the papier-mâché. Trees have always produced a material which is a composite — wood. Ordinary wood is a composite in which tubes of cllulose fibre are bonded by a natural polymer known as lignin (*Figure 1.18*). Composite materials are surprisingly (?) common.

Appendix 1.3 Alloys

An alloy is a mixture of a metal with other elements. Alloys have been used for more than 2000 years. Bronze was made from copper and tin and was one of the earliest alloys. It generally contained about one part of tin with nine parts of copper. By varying the relative proportions of the two metals, alloys with different properties can be made. The greater the amount of tin the harder the alloy, the greater the amount of copper the softer the alloy.

The coins you have in your pocket are examples of alloys — probably copper alloys. A coin made of pure copper would be very soft and easily damaged. A copper alloy coin is, however, much harder and thus not so easily damaged.

Coins	*Percentages by mass*			
	Copper	*Tin*	*Zinc*	*Nickel*
½p, 1p, 2p	97	0.5	2.5	—
5p, 10p, 50p	75	—	—	25

The metals used in motor cars are almost all alloys. The bodywork is generally mild steel, an alloy of iron with carbon and small amounts of other metals. The cylinder block might be an aluminium alloy or perhaps a magnesium alloy. The door handles might be a zinc alloy.

If you look around you at the wide variety of metals it is more than likely that all the metals you see will be alloys. Pure metals are relatively weak.

2 Structures

*Multi-storey office block, Hanley,
Stoke-on-Trent*

Objectives

The intention of this chapter is to relate the properties of materials to
their use in structures. It is assumed that Chapter 1 of this book and the
parts of *Book 1: Motion and Force* relating to force as a vector have
been covered.

The general objectives for this chapter are that after working through
it you should be able to:

(a) Explain what is meant by equilibrium;
(b) Use the conditions of equilibrium to determine forces acting on
     beams;
(c) Define the concept of moment;
(d) Explain for structures, such as an arch, how the properties of the
     material are being used to the best advantage.

Teaching note

Further reading can be found in *The New Science of Strong Materials*
and *Structures* both by J.E. Gordon. Experiments to investigate load-
ings in structures can be carried out using extension and compression
balances.

Equilibrium of structures

The photograph at the beginning of this chapter shows a building in the course of construction. The building is constructed on a steel frame. The steel columns are hollow, with a rectangular cross section of 300 mm by 300 mm. What are the requirements which determine which material to use and what size the columns should be?

The following is taken from a publication (*SHS the Builder*) by the manufacturer of the steel (British Steel Corporation, Tubes Division).

'Requirements for structural safety and serviceability.
For example, the structural framing must be sufficiently strong to transfer the vertical and horizontal loads to the foundations and to be sufficiently rigid to limit the vertical deflections and horizontal sway. Besides containing horizontal and vertical movements (perhaps brought about by differences in temperature and wind loading) the structural framing must also be durable and capable of providing adequate fire resistance.'

Systems which do not normally move when in use, e.g., buildings and bridges, are called **static systems.** The building shown in the photograph may look a rather complex system. The bridge shown in the photograph at the beginning of the book also looks a rather complex system. We can, however, consider the individual parts of the system, e.g., a single beam. The beam will be acted on by a number of forces – what is the behaviour of the beam? A vital point is whether the forces acting on the beam are such as to cause it to move. If the beam does not move then we say that it is in **static equilibrium.** This chapter looks at the conditions for such equilibrium.

Questions

1 In the quotation above the term 'rigid' is used.
What is meant by this term? What would you look for in a material if you wanted to find a rigid material?

2 How could differences in temperature bring about movements?

Will it move?

If an object is to remain at rest there should be no net force acting on it (Newton's first law of motion – see *Book 1: Motion and Force*). Forces add or subtract as vector quantities and thus due consideration has to be given to both their size and direction. Thus, to compute whether a particular object is in equilibrium we can use the triangle of forces (see *Book 1: Motion and Force*). If three forces acting at a point are represented in magnitude and direction by arrow-headed lines then these lines, when taken in the sequence that the forces act on the object, must form a triangle.

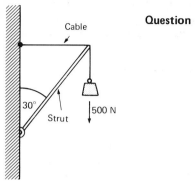

Figure 2.1

Question   3   *Figure 2.1* shows a strut pivoted at one end and held in position by a cable. If the strut is in equilibrium what must be the tension in the cable?

The cable holding the strut in place in *Figure 2.1* must be able to withstand the applied forces without yielding. The strut is under compression and it must withstand the compressive forces without buckling or yielding. The manufacturer of the steel columns used in the building shown in the opening photograph supplies 'safe load tables' which can be consulted when the steel sections are being considered for use. Thus for the 300 mm × 300 mm steel sections used in that building the following is part of the table.

Thickness/mm	Max. axial tension/kN	Max. axial compression (kN) for length (m)							
		0.5	1.0	1.5	2.0	2.5	3.0	3.5	4.0 etc.
10.0	1798	1783	1767	1752	1736	1721	1706	1690	1673
12.5	2216	2197	2178	2159	2140	2121	2102	2082	2061
16.0	2806	2781	2757	2732	2708	2683	2659	2634	2607

Hence, if the column was subject to a compressive force of 2000 kN and was 3.0 m long then the thickness of the 300 mm × 300 mm section that would be required would, as an absolute minimum, be 12.5 mm. Some factor of safety might well be allowed for as the calculation of the resultant compressive force on the column might not have taken into account all the various forces that might act on the column.

Will it tip — rotate?

Figure 2.2(a) shows a see-saw. What are the conditions for the see-saw to be in equilibrium and not to rotate or tip in one direction or the other?

Figure 2.2(b) shows the forces acting on the beam. There is a force due to each child; these forces act at the ends of the beam, where the children are sitting, and are in a direction at right angles to the surface of the earth. They are gravitational forces. The pivot is shown exerting a force in an upwards direction on the beam. There must be a force at the pivot as the beam is pressing the pivot into the ground and, if the pivot does not move, there will be no net force at that point. The total force exerted by the beam on the pivot is $F_1 + F_2$. (Neglecting the mass of the beam.) Thus the force exerted by the pivot on the beam must be opposite and equal in size to $F_1 + F_2$.

$$F_3 = F_1 + F_2$$

One condition for equilibrium is thus: there must be no resultant force in any direction.

The forces F_1 and F_2 create a turning effect about the pivot. The product of the force and the radius of its possible arc of rotation is termed the **moment** of the force. Thus, for force F_1 the moment about the pivot is $F_1 \times d_1$, and for force F_2 the moment about the pivot is

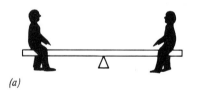

(a)

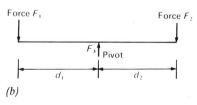

(b)

Figure 2.2 (a) The see-saw, (b) the see-saw force system

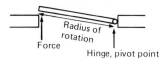

Figure 2.3

$F_2 \times d_2$. When you open a door you apply a moment to cause the door to rotate about its hinge — the moment is the product of the force you apply to the door handle and the radius of the possible arc of rotation (*Figure 2.3*).

Moment = force $\times$ radius of possible arc of rotation

Units: force – N, radius – m, moment – N m.

If you try an experiment with a pivoted beam and known forces at known distances from the pivot the reason for specifying this quantity moment as the product of force and radius becomes apparent. When the beam is in equilibrium, i.e., not rotating or tipping, the sum of the moments trying to cause a rotation in the clockwise direction equals the sum of the moments trying to cause a rotation in the anticlockwise direction. Thus, in the case of the see-saw:

Anticlockwise moment $= F_1 \times d_1$

Clockwise moment $= F_2 \times d_2$

Each moment is about the pivot point. At equilibrium

$$F_1 \times d_1 \; = \; F_2 \times d_2$$

A second condition for equilibrium is thus: the sum of the anticlockwise moments must equal the sum of the clockwise moments.

It does not matter about which point on a beam the moments are considered provided all the forces are taken into account. At equilibrium the anticlockwise moments about any point equal the clockwise moments about the same point. Thus, in the case of the see-saw, considering the moments about the end at which force F_1 is applied:

Clockwise moment $= F_2 \times (d_1 + d_2)$

Anticlockwise moment $= F_3 \times d_1$

It can be shown that this is the same as when moments are taken about the pivot by using the fact that, at equilibrium, $F_3 = F_1 + F_2$.

$$(F_1 + F_2) \times d_1 \; = \; F_2 \times (d_1 + d_2)$$

$$F_1 d_1 + F_2 d_1 \; = \; F_2 d_1 + F_2 d_2$$

$$F_1 d_1 \; = \; F_2 d_2$$

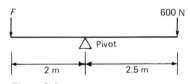

Figure 2.4

Questions

4   What force must be applied 2 m from the pivot point in *Figure 2.4* if the beam is to be in equilibrium? What is the reaction at the pivot? Neglect the mass of the beam.

5   *Figure 2.5* shows a beam supported at each end. It could be one of the floor beams supported on the outer columns in the photograph at the beginning of this chapter. Calculate the forces acting on each of the supporting columns, neglecting the mass of the beam.

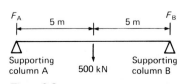

Figure 2.5

6   How would the forces on the columns in *Figure 2.5* change if an extra load of 200 kN acts a distance of 2 m from column A?

In the above problems the effect of the mass of the beam itself has been neglected. This, too, will contribute a force. If you try and balance a ruler across a pencil you will find that there is just one position of the ruler for which equilibrium occurs. The anti-clockwise moments due to the mass of the ruler on the left side of the pivot must be balanced by clockwise moments due to the mass on the right side of the pivot. The same balance condition would have been found if all the mass of the ruler had been concentrated at the centre of the ruler, directly over the pivot. This point on the pivot axis where the balance occurs and where all the mass of the ruler can be considered to be concentrated is called the **centre of mass**, sometimes referred to as the **centre of gravity**. For symmetrical homogeneous objects the centre of mass is the centre of symmetry of the object.

Question   7   How would the answers to questions **5** and **6** be changed if the beam had a mass per metre of 45 kg?

Building

If you wanted to bridge a small stream, you might well make the bridge out of planks of wood. A simple bridge consists of a single plank with its ends resting on the banks (*Figure 2.6*). The mass of the plank itself and the force produced by a person walking across the plank will both cause it to bend, the upper surface being in compression and the lower surface in tension. Wood is stronger in tension than in compression. For spruce, along the grain, the breaking stress in tension is 100 MN/m² and the crushing stress in compression is 30 MN/m². Thus the wooden bridge is most likely to fail first on its upper surface where the compressive stresses are experienced.

The ancient Greeks used stone beams to bridge gaps in their temples (*Figure 2.7*). The breaking stress in tension for the stone is probably

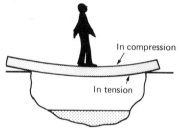

Figure 2.6 A simple wooden bridge

Figure 2.7 The Parthenon, Athens

about 4 MN/m² while the compressive crushing stress is about 500 MN/m². Stone is much stronger in compression than in tension. The enormous mass of the stone beam could be such as to exceed the breaking tensile stress and start cracks appearing. Such cracks would appear on the lower surface, that being the surface in tension.

The mild steel used for the steel beams in the building shown in the opening photograph of this chapter has a tensile breaking strength of about 550 MN/m². High tensile steel has a breaking strength in tension of as much as 1000 MN/m². A simple steel beam can be used to bridge a gap and will withstand much higher tensile stresses than the wood or the stone. The development of such materials, which are able to withstand high tensile stresses, is a comparatively modern development.

The open structure of the building shown at the beginning of this chapter would have been difficult to achieve without a material that withstood high tensile stresses — the floor/ceiling beams are just like the simple bridge in *Figure 2.6*. My home is an 'open-plan' house with a large area without any walls. Old houses had to have a large number of walls, and hence small rooms, because they could not support the ceilings/upper floors by beams alone. Wooden beams were just not strong enough, and they were generally not available in long lengths. Open-plan houses usually have a steel girder resting on the outer walls, and spanning the living area, to take the main load.

The structures considered for the timber, stone and steel are all of the same form — a beam supported at its ends. These and other forms of structure have to make use of the 'best' properties of a material. The arch is a structure which uses stone in compression, rather than tension, and so makes use of the strength of stone in compression (*Figure 2.8*).

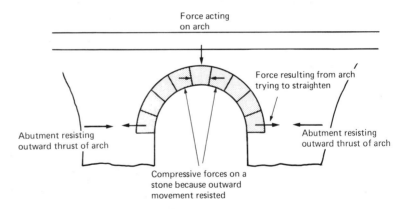

Figure 2.8 An arch

The vertical forces on the stones in an arch cause the stones in the arch to become compressed — the arch is trying to straighten out. To prevent the arch straightening out there must be some counterbalancing thrust from the abutments against which the arch pushes. These abutments are generally made fairly massive because of such forces.

The oldest surviving arched bridge is probably one at Smyma in Turkey and dates from the ninth century B.C. Roman bridges using arches still exist.

Figure 2.9 The nave of King's College Chapel, Cambridge

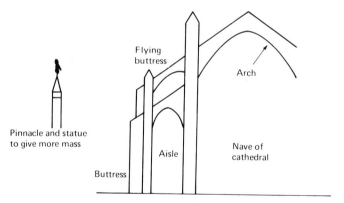

Figure 2.10

Arches were used for many structures. *Figure 2.9* shows the arches over the nave of a church. The arch is a good method of supporting the roof over a large open area. A problem with this arch is to prevent it straightening out and pushing the side walls of the church outwards. This thrust can be resisted by the use of buttresses (*Figure 2.10*). These sometimes have pinnacles and statues on top to give extra mass.

Figure 2.11 Reinforcement rods in concrete

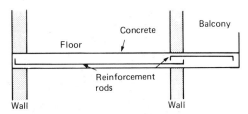

Figure 2.12

Many modern structures are built using reinforced or prestressed concrete (*Figure 2.11*). Concrete is weak in tension but strong in compression. Reinforcement of concrete by steel rods enables the concrete composite, called reinforced concrete, to carry greater tensile loads than would otherwise be possible. The stress is shared between the concrete and the steel rods. As steel costs more than concrete the steel rods are placed, generally, just in the regions where tensile stresses are likely to occur. Thus, in the case of a floor (*Figure 2.12*) which is supported at its extremities the rods are placed in the lower part of the floor thickness. In the case of a cantilevered balcony the rods would be placed in the upper part of the floor thickness. In prestressed concrete the steel rods are put permanently in tension, so putting the concrete in compression. This is done by holding the steel rods under tension while the concrete sets around them, the tension being released when the concrete has set.

Questions 8   A table can consist of just four legs and a top, the legs being at the four corners.

(a) What will be the effect of putting a large load on the table? Consider the effect on both the table top and the legs.
(b) Why do many tables have cross-pieces which link the legs together?

9   The photograph at the beginning of this book is of the Severn suspension bridge.

(a) Are the main suspension cables in tension or compression?
(b) Are the supporting towers in tension or compression?
(c) What would happen to the towers if the ends of the cables were not well anchored?

Suggestions for answers

1   Rigid means that the beam does not bend easily. This means the material has a high Young's modulus (see Chapter 1).

2   Expansion and contraction due to temperature changes could cause movements. If the movements were restrained the temperature change could produce stresses in the beams (see Chapter 1).

3   You could draw the triangle of forces to scale or use calculation. The tension is about 290 N.

4 $2F = 600 \times 2.5$, hence F is 750 N

5 $F_A + F_B = 500$ kN; taking moments about A gives $5 \times 500 = 10 F_B$. $F_B = 250$ kN, $F_A = 250$ kN. You might have obtained the answer just by considering the symmetry of the arrangement.

6 $F_A + F_B = 500 + 200 = 700$ kN; taking moments about A gives $2 \times 200 + 5 \times 500 = 10 F_B$. $F_B = 290$ kN, $F_A = 410$ kN.

7   The mass of the beam is 450 kg and so the gravitational force is about 4400 N. Hence, for question 5, $F_A = F_B = 252.2$ kN. In question 6, $F_B = 292.2$ kN, $F_A = 412.2$ kN.

8   (a) The upper surface of the table will be in compression and the lower in tension. This can have the effect of pulling in the top of the table legs and so result in the legs splaying out on the floor.
(b) To prevent the legs splaying out when loads are placed on the table.

9   (a) Tension
(b) Compression
(c) The tops would be pulled towards the centre of the bridge

---

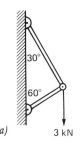

'a)   3 kN

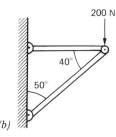

200 N

'b)

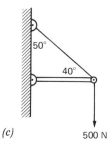

(c)   500 N

Figure 2.13

Further problems   No suggestions for answers are given for these problems.

10   Determine the forces in each member of the structures shown in *Figure 2.13*, neglecting the masses of the beams and cables.

11   A rectangular cross-section beam is to be used in a situation where it will experience a tensile force of up to 500 kN. The yield stress of the steel used for the beam is 300 MN/m^2.

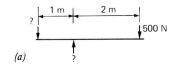

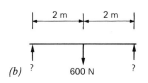

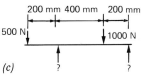

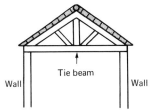

Figure 2.14

What cross-sectional area should be chosen for the beam if, for safety reasons, it should withstand four times the maximum stress for which it is designed?

12   A see-saw of mass 10 kg is balanced with a child at each end of the centrally pivoted plank.
     If each child has a mass of 40 kg what is the force exerted on the pivot by the plank?

13   Calculate the unknown forces in *Figure 2.14* if each beam is in equilibrium, and all beams are of negligible mass.

14   A beam of mass 20 kg per metre rests on two supports, one at each end, 6 m apart.
     What are the reactions at the end supports due to the mass of the beam?

15   Examine a chair and consider the forces acting on it during use. Some chairs have cross-pieces holding the legs together.
     Why? How does the property of the material used for a chair determine the form of the chair? Discuss.

16   Explain how the arch structure enables materials that are weak in tension to be used in situations where as a simple beam they might well fail.

17   Explain the purpose of the tie beams in the roof structure shown in *Figure 2.15*.

18   The crushing strength of brick is about 500 MN/m^2.
     If brick has a density of 2500 kg/m^3 how high a wall could be built before the bottom brick was crushed?

19   The previous question indicates that very high walls should be possible. Walls, however, frequently tumble down without ever getting anywhere near the maximum height possible.
     Why do walls tumble down easily?

Figure 2.15

Appendix 2.1   Articulated tower-crane jibs

A product of elementary logic

Like most good ideas, as Henry Foster explains, the articulated tower-crane jib is so simple.

Prominent today on the urban skyline is the articulated tower-crane jib. This invention is so useful to the construction industry – and articulation is based on such simple principles – that one wonders why it was not conceived long ago. Briefly, it gives a 25 per cent increase in load capacity at a substantially lower capital cost.

To understand the importance of articulation it is necessary to go back to square one and define the stresses that are imposed on the jib and mast of a 'conventional' tower crane.

The mast of any tower crane must bear its own weight, the weight of the jib under load and considerable wind stresses which can amount to as much as one-third of the total strain on the mast. The jib is a metal arm held under stress to the mast by a steel restraining rope which, with the counter weight, keeps the crane in balance under load. On conventional cranes there is only one restraining rope or tie bar (*Figure 2.16(a)*).

Tower-crane manufacturers, when designing machines with greater capacities and greater outreach, build the jib from heavier steel sections. This increases the dead-weight that has to be carried by the mast so that, too, has to be built from heavier material. Inevitably, this puts up the price.

The advent of industrialised building systems, using heavy pre-cast panels, has led to a world demand for tower cranes with longer jibs capable of taking bigger loads. There seems no end to this demand, and the tower cranes of the future will undoubtedly have to be very much larger machines than anything on the market today.

If these super-cranes were built on traditional lines — using heavier steel members — they would sell at a prohibitively high price. The tower-crane industry has been aware of this problem for some years, and a number of manufacturers have been experimenting with designs that would enable the capacities and outreach of their machines to be increased without any compensating increase in the weight of the steel structure.

The main difficulty was to overcome the 'bending moment' of the crane when under maximum load at maximum outreach. A tower crane is a pliable, supple structure which bends under load in much the same manner as a longbow (*Figure 2.16(b)*). The longer the jib, the greater its tendency to bend. Thus, if the jib were lengthened without being strengthened, the crane would be totally unsafe.

One of the early solutions was to fix two restraining ropes to the jib (*Figure 2.16(c)*) to support the greater length. Theoretically it is the ideal answer, but in practice the cure is worse than the disease.

The two restraining ropes obviously cannot be of the same length, one must be considerably longer than the other. They will not, therefore, expand and contract, following changes in temperature, to the same degree. Thus, they cannot support the jib in the way in which the designer intended. Inevitably the jib is restrained unequally by the two ropes, causing it to develop ridges along its main apex members (*Figure 2.16(c1)*). This, in turn, overstresses other parts of the crane and defeats the whole purpose of the structural design.

It can be argued that a computer-based design would avoid most of these difficulties. However, it would involve precision engineering of a very high order making any such tower-crane a most expensive — and commercially unattractive — proposition, presenting considerable manufacturing problems.

The twin-rope idea did not catch on, and the next experiment was to fix six or more restraining ropes along the entire jib length. However, differences in tension caused a whole string of ridges along the apex (*Figure 2.16(d)*).

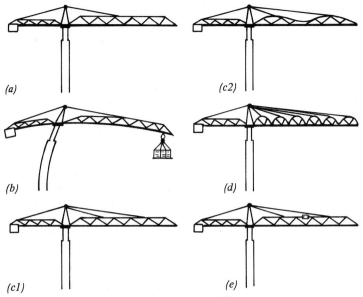

Figure 2.16 The development of the articulated jib

Pierre Pingon, who designed the tower-crane range that bears his name, presented the obvious solution: split the jib into two independent sections, tied to the mast by two independent steel ropes. This, in fact, is articulation. There are two comparatively short jib-lengths held in place by two independently suspended tie-ropes (*Figure 2.16(e)*). This is a principle similar to that used by suspension-bridge engineers for the past three or four generations — and engineering-wise the problems that confronted tower-crane designers were much the same as those that faced bridge builders. Both have to take a load in suspension.

Take a closer look at *Figure 2.16(e)*. The two jib sections held together by steel tie-ropes can move independently. Since each tie-rope is responsible only for its own 'section', it makes no difference whatsoever whether the two ropes are in perfect tension or not. You have two jibs for the price of one.

With this system one can either increase the jib length or its carrying capacity without any substantial increase in the overall weight of the jib structure. If this overall weight is not greatly increased, then it follows that the mast members can be considerably lighter than those on a 'standard' tower-crane of equivalent capacity and outreach. Thus, articulation enables the load or outreach of the crane to be increased without adding an inordinate amount to manufacturing costs.

This article by H. Foster appeared in *Project*, Autumn 1968 and was an adaptation of an article which had appeared previously in *600 Magazine* in 1965.

3 Fluids

The new Claerwen dam

Objectives

The intention of this chapter is to introduce, or most likely revise, concepts such as density, pressure and expansion and apply them to both liquids and gases. It is assumed that you can handle indices.

The general objectives for this chapter are that after working through it you should be able to:

(a) Discuss the meaning of the term fluid;
(b) Solve problems involving density and relative density;
(c) Solve problems involving pressure and its measurement by means of manometers;
(d) Explain a system, such as a hydraulic braking system, based on the transmission of pressure;
(e) Explain Archimedes' principle and use it in problems involving flotation;
(f) Define the coefficients of apparent and real expansion for liquids and solve problems using these coefficients;
(g) Describe the anomalous behaviour of water to variations of temperature;
(h) State the ideal gas laws and solve problems using them;
(i) Recognise the problem of specifying a scale of temperature;
(j) Recognise the limitations of the gas laws.

Teaching note

Experiments appropriate to this chapter can be found in *Nuffield Physics: Pupils' Text Year 3* and *Project Physics Course, Handbook*.

What are fluids?

Fluids are substances that flow. We can divide all materials into two sets: those that flow and those that do not. If you put a block of steel on the table top you expect it to retain its shape and position on the table, i.e., not to flow. If you put a 'block' of water on the table top you would be surprised if it retained its shape, it flows over the table. Similarly if you emptied a container of gas, say carbon dioxide, on to the table top you would be surprised if it remained as a 'block' of gas, it flows. Liquids and gases flow, solids do not.

The distinction between the behaviour of solids and the 'rest' appears clear cut. In practice it is not so clear. Solids do flow — if you wait long enough. Some ancient lead roofs are measurably thicker at the eaves than at the roof apex as a result of the lead flowing very slowly over the centuries.

A simple distinction between solids and the 'rest' is that solids do not need containers to remain together in a 'pile' or 'block', fluids do need containers. Here again the distinction is not always clear. A drop of mercury will keep 'together' as a drop but a fine powder is liable to flow all over the place.

Questions

1 You can buy paints which are known as 'non-drip'. The paint is fluid when applied rapidly with a brush — it can be spread over a surface — but if left on a surface it does not run. Is the paint a fluid?

2 Silicone putty, known generally as 'potty putty', flows like a liquid if put on the table. It can however be moulded into a ball shape and if thrown against a wall will bounce off. If the ball is hit with a hammer it fractures. Is the putty a solid?

Density

The following table shows typical densities of some liquids and gases, density being mass per unit volume. The density of liquids, as for solids, depends not only on the material concerned but also on temperature, hence temperatures are quoted for the values given. The density of a gas depends on both the temperature and the pressure and so the values are quoted for normal room temperature and pressure conditions.

Material	State	Temperature	Density/kg m^{-3}
Water	Liquid	18°C	999
		4°C	1000
Mercury	Liquid	18°C	13600
Ethyl alcohol	Liquid	18°C	810
Air	Gas	Room	1.2
Carbon dioxide	Gas	Room	1.3
Hydrogen	Gas	Room	0.09

Questions **3**   What is the mass of 200 cm³ of water at room temperature, 18 °C?

4   What is the mass of the air in a typical room?

The term **relative density** (sometimes called **specific gravity**) is sometimes used and is the density of a material relative to that of water (generally taken as 1000 kg/m³).

$$\text{Relative density} = \frac{\text{density of material}}{\text{density of water}}$$

Thus the relative density of mercury is 13.6. There are no units for relative density as it is a ratio.

Questions **5**   What is the relative density of ethyl alcohol?

6   What is the relative density of steel? (Refer to Chapter 1 for the density of steel.)

Pressure

If you press a drawing pin into a board you soon know whether or not you are pressing on the right end of the drawing pin. When the force is spread over the cap of the drawing pin there is no problem. When the same force is spread only over the point of the pin there is a problem — it hurts! The area over which a force is applied is important. The force applied at right angles to an area divided by that area is called **pressure**.

$$\text{Pressure} = \frac{\text{force}}{\text{area}}$$

Units: force – N, area – m², pressure – N/m² or N m⁻² or pascal (Pa). (1 Pa = 1 N/m².)

Questions **7**   If you push on the drawing pin with a force of 100 N what is the pressure exerted by the point of the pin on the surface into which it is applied if the point has a diameter of 0.05 mm (area about 0.002 mm²)?

8   Liquids can exert pressure. Consider the force acting on the base of a beaker due to the mass of liquid above it in the beaker (*Figure 3.1*).

(a) What is the volume of the liquid in the beaker?
(b) What is the mass of the liquid in the beaker?
(c) What is the gravitational force acting on the mass of the liquid?
(d) This gravitational force is acting on the base of the beaker. What is the force per unit area of beaker base? What is the pressure exerted by the liquid on the base of the beaker?

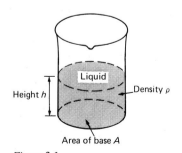

Figure 3.1

(e) Your answer to (d) should be

$$p = h\rho g$$

The answer is independent of the cross-sectional area of the beaker.

What is the pressure on the base of a beaker if it contains water to a height of 100 mm above the base?

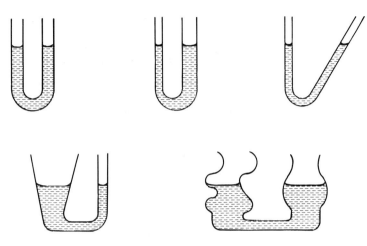

Figure 3.2

Figure 3.2 shows a number of U-tubes. Each arm of the U contains the same liquid. The vertical height of the liquid in each arm of any U-tube is the same. Thus the pressure at the base of each limb of a U is the same. This applies irrespective of the form or diameter of the different limbs of a U, the pressure due to the liquid depends only on the height of the liquid, and is given by $p = h\rho g$.

If you poured liquid into one of the U-tubes you might for some instant of time have the liquid higher in one limb of the U than the other. The liquid will, however, quickly flow from the limb where it is higher to the other limb until the height of liquid in each limb is the same. The fact that the liquid flows between the limbs means that there must be a force component acting in the direction of the movement. The fluid flows until there is no pressure difference. The pressure, calculated as acting on the base of the U, due to the height of liquid above, must therefore exert forces in all directions not just on the base. Pressure acts perpendicularly to any surface placed in a fluid, irrespective of its direction — there is thus no reason to specify direction when talking of pressure but refer simply to the size of the pressure at a point.

If you take a can with holes in the sides and fill it with water then the water comes spurting out of the holes (*Figure 3.3*). The fact that the water comes spurting out of holes in the sides of the can shows that pressure acts in all directions. But there is also another point to observe — when the water starts to come out of a hole it comes out at about right angles to the surface of the can. The force exerted on a surface by the liquid pressure is at right angles to the surface when the liquid is at rest.

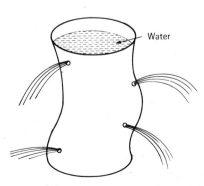

Water

Figure 3.3

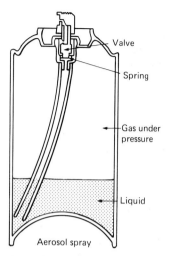

Figure 3.4 Aerosol spray

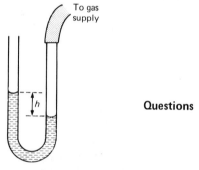

Figure 3.5 A simple manometer

Question   9   The photograph at the beginning of this chapter shows a dam holding back a large quantity of water.
Where would you expect the forces exerted by the water on the dam to be greatest? The question might well be phrased as — where would you make the concrete the thickest if you were designing a dam?

Like liquids, gases also exert pressures. The air pressure in a car's tyres manages to support the car. The aerosol spray contains gas under pressure (*Figure 3.4*), it is the pressure that forces the liquid out in a spray.

The pressure of a gas, such as that from the domestic gas supply, can be compared with the pressure due to a column of liquid by means of a U-tube containing the liquid (*Figure 3.5*). Such a U-tube is called a **manometer**. Before the gas enters the limb of the U-tube the liquid levels in the two limbs are the same. When the gas is admitted to one limb a difference in liquid level is produced. As the pressure at the base of each of the limbs of the U-tube must be the same then the difference in levels must be due to the difference in the pressure on the liquid surface of each limb. The gas pressure is thus $h\rho g$ (where h is the vertical height difference in levels and ρ the density of the liquid in the manometer) above that exerted by the atmosphere on the liquid level in the other arm of the U-tube. Due consideration has always to be given to the fact that experiments at the surface of the Earth are subject to a pressure due to the mass of air above us.

Questions   10   When a manometer containing water is connected to a gas supply the level of the water in the limb directly connected to the gas supply is depressed to 200 mm below that in the other limb which is open to the atmosphere.
What is the pressure difference between that of the gas and that of the atmosphere? Is the gas pressure higher or lower than the atmospheric pressure?

11   One of the limbs of a manometer containing mercury is connected to a vacuum pump which removes most of the air from that limb. The difference in level between the mercury in the two limbs is 760 mm, the mercury being highest in the limb connected to the pump.
What is the atmospheric pressure?

The **atmospheric pressure** at the surface of the earth is generally about 100 kN/m², i.e., 10^5 N/m² or 10^5 Pa. Standard pressure or one atmosphere pressure is defined as the pressure 1.01325×10^5 N/m². This is equivalent to a difference in level of 760 mm of mercury. The instruments used for the measurement of atmospheric pressure are called **barometers**. *Figure 3.6* shows a simple barometer. This is essen-

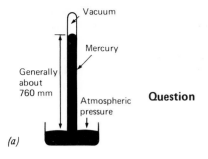

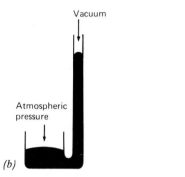

Figure 3.6 (a) The simple barometer, (b) the equivalent U-tube

tially a U-tube with one of the limbs of the U having a different diameter to the other. The barometer has what is virtually a vacuum above the mercury surface in the tube.

Question

12 How high would the water level be if water was used for the liquid in a simple barometer instead of mercury?

The transmission of pressure

If the pressure applied to the surface of a liquid is increased then the pressure everywhere within that liquid is increased by the same amount. Thus, if we consider a beaker of liquid, as in *Figure 3.2*, then the pressure at the base of the beaker due to the liquid is $h\rho g$ and the total pressure is the pressure due to the atmosphere acting on the surface of the liquid plus $h\rho g$. If the pressure acting on the surface is increased then the total pressure at the base of the beaker is increased by the same amount. The pressure change at the surface has been transmitted through the liquid.

The hydraulic braking system of a car depends on the transmission of pressure through a liquid, the brake fluid. *Figure 3.7* shows the basic arrangement. Pressure is applied by the driver's foot pressing the brake pedal and this pressure is transmitted to the brake shoes, causing them to be pressed against the brake drums.

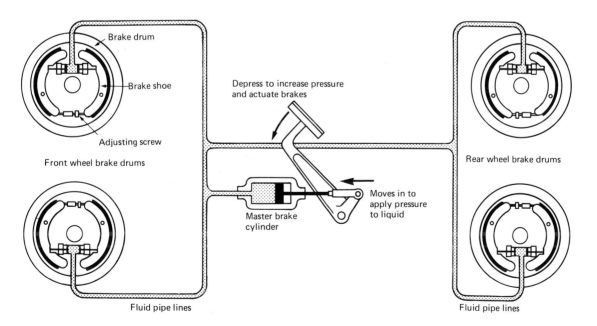

Figure 3.7 Hydraulic motor-car brake system

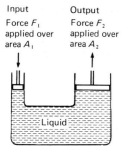

Input
Force F_1
applied over
area A_1

Output
Force F_2
applied over
area A_2

Liquid

$F_1/A_1 = F_2/A_2$
$F_2 = F_1 A_2/A_1$
If A_2 is larger than A_1
then F_2 is larger than F_1

Figure 3.8

The transmission of pressure by a liquid is used in many machines, e.g., the hydraulic jack, hydraulic press, transmission systems in tipping lorries, etc. In many of these, because it is pressure that is transmitted undiminished, large forces can be generated as a result of the application of much smaller forces. The pressure change can be produced by applying a force over a small area, the output appearing as a larger force over a larger area, the pressure in each instance being the same. *Figure 3.8* shows the general principle.

Question   13   If in *Figure 3.8* the smaller piston has a cross-sectional area of 100 mm² and a force of 1 N is applied to it, what will be the force acting on the larger piston of area 10 000 mm²?

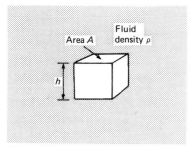

Area A

Fluid
density ρ

h

Figure 3.9

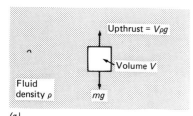

Upthrust = $V\rho g$

Volume V

Fluid
density ρ

mg

(a)

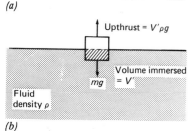

Upthrust = $V'\rho g$

Volume immersed
= V'

mg

Fluid
density ρ

(b)

Figure 3.10 (a) An object floating completely immersed in a fluid, $V\rho g = mg$, (b) an object floating partially immersed in a fluid $V'\rho g = mg$

Archimedes' principle

An object immersed in a fluid must be acted on by the pressures in that fluid. The upper surface of an object is, however, at a different depth to its lower surface (*Figure 3.9*). Thus the pressure on the upper surface must be less than that on its lower surface, less by an amount related to the difference in depths of the two surfaces concerned. If this is a vertical height difference of h then the pressure difference must be $h\rho g$, where ρ is the density of the fluid. This pressure difference means that there must be a difference in the forces acting on the upper and lower surfaces. The net force acting on the lower surface must be greater than that acting on the upper surface by $h\rho gA$. But hA is the volume of the solid or the volume of fluid displaced to make way for the solid. The mass of this displaced fluid is thus $hA\rho$ and thus $hA\rho g$ must be equal in size to the gravitational force acting on the mass of fluid displaced. This net force is acting in an upwards direction and is thus known as the **upthrust**. The upthrust is equal to the weight of the displaced fluid. This is known as **Archimedes' principle**.

An object floats when there is no net force acting on it (*Figure 3.10*). This means that the upthrust must be equal in size to the gravitational force acting on the object. A floating body must therefore displace its own weight of fluid. An object only partially immersed in a liquid is displacing only part of its volume of liquid, that part of the volume below the surface of the liquid. This volume of liquid has, however, sufficient weight to be equal in size to the gravitational force acting on the entire object.

Questions   14   Calculate the upthrust acting on an object having a volume of 10^{-6} m³ and a density of 7800 kg/m³ when immersed in water of density 1000 kg/m³.

15   Calculate the upthrust acting on the object in question **14** when it is in air of density 1.2 kg/m³.

16   A balloon has a volume of 0.008 m^3 and contains hydrogen of density 0.09 kg/m^3.
     What is the net force acting on the balloon if it is in air of density 1.2 kg/m^3? Assume that the fabric of the balloon has a mass of 3 g.

17   Ice has a density of 917 kg/m^3.
     What fraction of the volume of a cube of ice will be below the surface when an ice cube floats in water of density 1000 kg/m^3?

18   What can you say about the density of an object if it floats in a liquid?

A block of iron is more dense than the corresponding volume of water and thus the upthrust on a block of iron in water is not as large as the gravitational force acting on the iron. The iron thus sinks when placed in water. Ships of iron, however, do not sink. This is because they are not solid blocks of iron. The weight of water displaced by the volume of the ship below the water line is equal to the total gravitational force on the ship due to the mass of its structure and the load carried. When a ship is carrying an extra load it floats deeper in the water. Special lines, called **Plimsoll lines**, are marked on the sides of ships to show to what depth they can sink when loaded.

Questions

19   How do lifebelts enable a person to float in water?

20   A lifebelt has a volume of 0.020 m^3 and when worn by a man of mass 70 kg supports him with two tenths of his volume above water, the belt being completely immersed. The sea water has a density of 1100 kg/m^3 and the average density of the man is 1200 kg/m^3.
     What is the average density of the lifebelt?

The **hydrometer** is an instrument used for the measurement of the density of a liquid. When the hydrometer is placed in the liquid (*Figure 3.11*) it sinks to a depth which depends on the density of the liquid concerned and the weight of the hydrometer concerned. The stem of the instrument has a scale marked in density, or relative density, which has been obtained by calibration. The greater the density of the liquid in which the hydrometer is placed the smaller the depth to which it sinks. Hydrometers which can be used for different density ranges can be produced by having different masses in the base of the hydrometer.

If V is the total volume of the hydrometer up to the zero point on the scale on the stem and A the cross-sectional area of the stem, then if the hydrometer floats with a length L above the zero scale point immersed

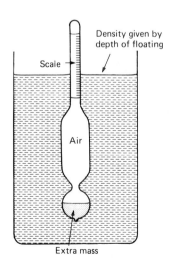

Figure 3.11 A hydrometer

$$(V + AL)\rho g = mg$$

where ρ is the liquid density and m the mass of the hydrometer. This can be simplified to

$$\frac{V + AL}{m} = \frac{1}{\rho}$$

$$\frac{V}{m} + \frac{AL}{m} = \frac{1}{\rho}$$

or

$$K + CL = \frac{1}{\rho}$$

where K and C are constants for the hydrometer concerned.

Expansion of liquids

Liquids expand when their temperature rises. The simple mercury-in-glass thermometer depends on this to give a measure of temperature. When a liquid is in a container and the temperature rises both the container and the liquid expand. Thus, if the container had graduations and we made no correction for its own expansion we would read the apparent expansion of the liquid. The apparent expansion is found to depend, to a reasonable approximation, on the initial volume of the liquid, the change in temperature and the materials of both the liquid and the container. This relationship can be expressed as

Change in volume = initial volume × change in temperature
× coefficient of apparent expansion

$$V_\theta - V_0 = V_0 \times \theta \times \gamma$$

where V_0 is the initial volume, V_θ the volume after a temperature change of θ and γ the **coefficient of apparent expansion**.

Units: $V - \text{m}^3, \theta - {}^\circ\text{C}, \gamma - {}^\circ\text{C}^{-1}$.

The volume given by the apparent coefficient is lower than the real amount by which the liquid expands because the vessel containing the liquid has also expanded. The real volume change of the liquid after a temperature change of θ equals the apparent volume change plus the amount by which the vessel has expanded.

$$\underset{\text{real}}{(V_\theta - V_0)} = \underset{\text{apparent}}{(V_\theta - V_0)} + \underset{\text{container}}{(V_\theta - V_0)}$$

or

$$\gamma_{\text{real}} = \gamma_{\text{app}} + \gamma_{\text{container}}$$

The coefficient of cubical expansion of pyrex glass is $9 \times 10^{-6}\ {}^\circ\text{C}^{-1}$, thus the coefficient of real expansion of a liquid in a pyrex glass container is $9 \times 10^{-6}\ {}^\circ\text{C}^{-1}$ larger than the coefficient of apparent expansion of the liquid. The following are some typical coefficients of real expansion of liquids.

Liquid	Coefficient of real expansion/$°C^{-1}$
Water	210×10^{-6}
Mercury	180×10^{-6}
Ethyl alcohol	1080×10^{-6}
Benzene	1220×10^{-6}

Questions   21   Which will expand more when the temperature rises – a given volume of mercury in a glass beaker or the same volume of ethyl alcohol in an identical glass beaker?

22   What is the apparent coefficient of expansion of mercury in a pyrex container?

23   What will be the apparent change in volume of 200 cm^3 of water in a pyrex beaker when the temperature changes by 30 $°C$?

24   (a) What will be the real volume occupied by 1 m^3 of water when the temperature rises by 10 $°C$?
(b) The mass of the water in that volume will not have changed as a result of the temperature change. How will the density have changed? What will be the new density if the original density was 1000 kg/m^3?

When the temperature of a liquid, or a solid, rises the density decreases. This is because the volume occupied by a certain mass of the substance has increased.

$$V_\theta - V_0 = V_0 \theta \gamma$$

Hence

$$\frac{V_\theta - V_0}{V_0} = \frac{V_0 \theta \gamma}{V_0}$$

$$\frac{V_\theta}{V_0} - 1 = \theta \gamma$$

$$\frac{V_\theta}{V_0} = \theta \gamma + 1$$

But the mass of the liquid is given by

$$\text{mass} = V_0 \rho_0 = V_\theta \rho_\theta$$

where ρ_0 is the original density and ρ_θ the density after the temperature change of θ.
Hence

$$\frac{V_\theta}{V_0} = \frac{\rho_0}{\rho_\theta}$$

and so

$$\frac{\rho_0}{\rho_\theta} = \theta \gamma + 1$$

This relationship is often written as

$$\rho_0 = \rho_\theta (1 + \gamma \theta)$$

Figure 3.12 shows how the density of water changes with temperature. Water has an anomalous density variation with temperature. The density of water increases when the temperature rises from 0 °C to 4 °C, reaching a maximum at that temperature. Above 4 °C water starts to behave 'normally' — the density decreasing with an increase in temperature. This means that the coefficient of expansion of water is not constant — water has a negative coefficient of expansion between 0 °C and 4 °C — that is, it doesn't expand it contracts. The coefficients of expansion quoted earlier are only averages over a temperature range, materials do not expand uniformly with temperature over large ranges of temperature.

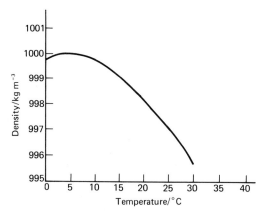

Figure 3.12 *The variation of the density of water with temperature*

Question 25 If you heat a beaker of water, or indeed any liquid, the water close to the base of the beaker becomes hotter than that near the surface.
What will happen to the density of the water near the bottom of the beaker? What will happen to the water near the bottom of the beaker as a result of the density change?

The behaviour of gases

The volume occupied by a mass of a substance, whether it is a solid, liquid or gas, depends on the pressure to which the substance is subject. The larger the pressure the smaller the volume. For solids and liquids the effect of increasing the pressure above atmospheric pressure is small enough to be insignificant for most situations, the volume barely changes for the pressure changes normally encountered. For a gas, however, the changes can be very noticeable.

If you take a bicycle pump and put your finger over the nozzle end when the pump handle is fully extended then when you come to push the handle in you 'squash' the air trapped within the pump (*Figure 3.13*).

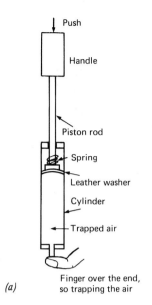

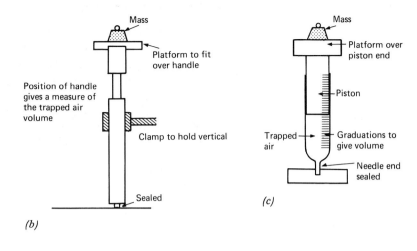

(b)

Figure 3.13 Investigating pressure/volume relationships for a gas

The pressure acting on the air in the pump is being increased. The more force you apply to the pump handle the further you can push the handle and the smaller the volume occupied by the air in the pump. The force you are applying acts over the piston area. *Figure 3.13(b)* shows how the arrangement can be modified to enable the pressure acting on the air and its volume to be monitored. *Figure 3.13(c)* shows how a syringe can be used for the same experiment.

The results of such an experiment show that to halve the volume the pressure acting on the air has to be doubled. To reduce the volume to one-third requires treble the pressure. *Figure 3.14* shows these results graphically. The pressure required is proportional to the reciprocal of the volume. This is sometimes stated as — the pressure is inversely proportional to the volume.

Figure 3.14 Pressure/volume relationships for a gas

$$\text{Pressure} \propto \frac{1}{\text{volume}}$$

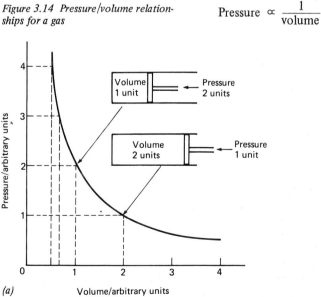

(a)

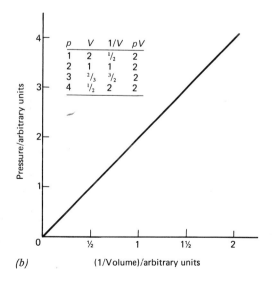

p	V	$1/V$	pV
1	2	$\frac{1}{2}$	2
2	1	1	2
3	$\frac{2}{3}$	$\frac{3}{2}$	2
4	$\frac{1}{2}$	2	2

(b)

This can be written as

Pressure $\times$ volume $=$ a constant

$pV =$ a constant

This relationship is known as **Boyle's law**, having been discovered by him in about 1660. The relationship only holds if the temperature and the mass of gas do not change during the experiment and not for all gases at all temperatures. For air at temperatures in the region of room temperature the law is, to a good approximation, obeyed.

Questions   26   A compressed air cylinder contains air at a pressure of 20×10^5 N/m^2. This is twenty times the atmospheric pressure. What would be the volume occupied by the air at 1×10^5 N/m^2? This is atmospheric pressure. Assume Boyle's law is obeyed.

27   A pump is used to pump air into a tyre. If the tyre pressure is at 1.8×10^5 N/m^2 and the pump takes in air at atmospheric pressure, 1.0×10^5 N/m^2, and compresses it in its cylinder, at what point in the stroke of the pump will air enter the tyre? Assume Boyle's law is obeyed.

About the year 1787 J.A. Charles discovered a relationship between the volume of a gas and its temperature, the gas pressure being kept constant. He did not publish his results and it was left to J.L. Gay-Lussac, who repeated the experiments, to express the relationship. The relationship discovered, generally known as **Charles' law**, was that the change in volume of the gases tested was proportional to the change in temperature. We can express this gas expansion in the same way as we expressed the expansion of a liquid or solid,

Change in volume $=$ initial volume $\times$ change in temperature $\times$ coefficient of expansion

Different liquids and different solids have different coefficients of expansion — with the so-called permanent gases, e.g., nitrogen, oxygen, air, there only appears to be one coefficient, the same for all the gases.

A graph of the volume of a permanent gas, at constant pressure and with constant mass, against temperature looks like *Figure 3.15(a)*. Because all these gases have the same coefficient of expansion we do not need to state the gas concerned. Thus if one of the gases was found to have a volume of 2.0×10^{-3} m^3 at 20 °C then at 47 °C the volume would be 2.2×10^{-3} m^3, regardless of which gas was being considered. By shifting the axes of the graph in *Figure 3.15(a)* we can produce the graph in *Figure 3.15(b)*. This means a different scale for temperature but does produce the great simplification that the volume V is directly proportional to the temperature T on this scale, i.e.,

$V/T =$ a constant

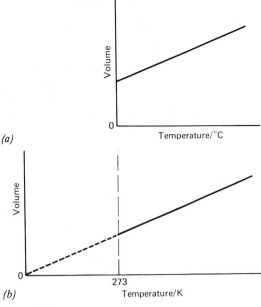

(a)

(b)

Figure 3.15

This scale was originally called the **Absolute scale** of temperature. The scale was defined in terms of the behaviour of what has been called an **ideal gas**. An ideal gas is one that obeys the above relationship exactly. The term **Kelvin scale** is now used (although it is defined in a different way). The temperatures on the Absolute scale, or the Kelvin scale, can be given by adding 273 to the temperatures on the Celsius scale (for an accuracy to only three figures). Temperatures on the Kelvin scale are indicated by the letter K appearing after the number denoting the temperature. Room temperature, about 18 °C, is thus about 291 K.

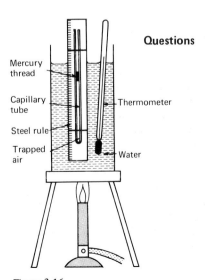

Figure 3.16

Questions

28 What are the temperatures on the Kelvin scale of (a) 20 °C, (b) −20 °C, (c) 200 °C?

29 To what temperature would a gas have to be raised for its volume to become twice that of its volume at 0 °C, the pressure and mass of the gas remaining constant?

30 A simple way of finding out how the volume of a gas such as air changes with temperature, as given by a mercury-in-glass thermometer, is to trap air in a tube with a thread of mercury (*Figure 3.16*), the other end of the tube being sealed. The tube is then stood in a beaker of water, with a thermometer, and the length of the air column measured at different temperatures.

(a) How would the length of the trapped air column and the cross-sectional area of the tube affect the results?
(b) Would a short length of mercury be better or worse than a long thread?

(c) Does the tube have to be kept vertical during the experiment or can it just be stood in the beaker and its position varied during the experiment?

(d) What type of relationship would you expect between the length of the air column and the temperature in °C?

Note: because the Absolute temperature scale is based on the expansion of a gas it would be better to consider the above experiment as determining how the mercury in the thermometer expands with temperature, the temperature being given by the expansion of the air.

31   When dealing with volumes of gases, in say chemical reactions, it is customary to refer them to a **standard temperature** of 273 K so that comparisons can be made more easily.

(a) A gas produced in a reaction is found to have a volume of 50 cm³ at 39 °C. What is its volume at the standard temperature?

(b) A reaction should produce 100 cm³ of a gas at the standard temperature. Because of a temperature change occurring during the reaction the gas will be produced at 100 °C. What will be the volume of gas produced at this temperature?

If a gas is enclosed in a constant volume enclosure then when the temperature rises it is not possible for the volume of the gas to increase. The increase in temperature produces an increase in pressure. For the so-called permanent gases

Change in pressure = initial pressure X change in temperature
X coefficient of expansion

The gases all have the same coefficient of expansion, the same as that for the volume expansion. We can thus write

$$p/T = \text{a constant}$$

where T is the temperature on the Kelvin scale. This means that the pressure of the gas, at constant volume and constant mass, is directly proportional to the temperature on the Kelvin scale.

32   Gas is stored in a cylinder at a pressure of 3 X 10⁵ N/m² and a temperature of 27 °C. What will be the pressure if the volume does not change but the temperature increases to 77 °C?

33   Sometimes at a party a balloon bursts, without being touched by anybody. Suppose a balloon was inflated to 9/10ths of the pressure at which it would burst. When the balloon was inflated the temperature was 18 °C. At what temperature would you estimate that the balloon would burst if no change in volume of the balloon were to occur?

For the so-called permanent gases the relationships describing their behaviour can be summarised as:

pV = a constant if temperature and mass are constant;
V/T = a constant if pressure and mass are constant;
p/T = a constant if volume and mass are constant.

These relationships can be combined into one equation

pV/T = a constant for a constant mass of gas

Question   34   A car engine takes air into the cylinders at atmospheric pressure 1×10^5 N/m² and a temperature of 27 °C. The compression ratio is 10 to 1, i.e., the volume is reduced by a factor of ten. This results in a pressure rise to 25 times the initial pressure. What is the resulting temperature of the air?

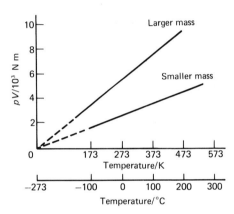

Figure 3.17

Figure 3.17 shows how pV for a gas depends on the temperature. The graphs are straight lines passing through the point – temperature (K) equal to zero, pV equal to zero. The slope of the graph is pV/T and depends on the mass of the gas concerned. It also depends on the particular gas concerned. The constant is not the same for 2 g of hydrogen and 2 g of oxygen. The equation can thus be written as

$pV/T = mR$

where m represents the mass and R is a constant which depends on the gas concerned.

Question   35   At normal atmospheric pressure of 1.013×10^5 N/m² and at a temperature of 20 °C a mass of 1 kg of dry air occupies a volume of 0.0831 m³. What is the value of the constant R in the above expression?

The following are some typical values for the constant R, generally called the **specific or characteristic gas constant**.

Gas	Characteristic gas constant/$\text{N m kg}^{-1}\text{ K}^{-1}$
Hydrogen	4160
Nitrogen	297
Oxygen	260
Dry air	287

The characteristic gas constant for hydrogen is 14 times that of nitrogen, 16 times that of oxygen. If you have studied chemistry these numbers may be familiar to you. The numbers represent the relative molecular masses of the gases concerned. The term **molar mass** is used for the sum of the atomic masses in a molecule of a substance when expressed relative to an atom of a particular isotope of carbon which has been given the value of precisely 12 g.

Gas	Molar mass/g
Hydrogen H_2	2.02
Nitrogen N_2	28.02
Oxygen O_2	32.00
Dry air, a mixture	28.96

The specific gas constant R can, thus, be related to some **universal gas constant** R_0.

$$\text{Specific gas constant}, R = \frac{\text{universal gas constant}, R_0}{\text{molar mass}}$$

The universal gas constant has the value $8.3143 \text{ N m mol}^{-1}\text{ K}^{-1}$. Thus for helium, which has a molar mass of 4.00 g, the specific gas constant is $8.3143/(4.00 \times 10^{-3})$ or about $2079 \text{ N m kg}^{-1}\text{ K}^{-1}$.

Questions

36   A gas cylinder holds $50 \times 10^{-3} \text{ m}^3$ of nitrogen. When full the pressure gauge on the cylinder reads $20 \times 10^5 \text{ N/m}^2$, the temperature being 17 °C. The pressure gauge indicates by how much the gas pressure is above atmospheric pressure. Atmospheric pressure is $1 \times 10^5 \text{ N/m}^2$

(a) What is the mass of gas in the cylinder?
(b) After some use the pressure reading has dropped to $5 \times 10^5 \text{ N/m}^2$, at the same temperature as the initial reading. What is the mass of the gas that has been used?

37   At 10 °C the density of carbon dioxide CO_2 is 1.90 kg/m^3 when the pressure is $1 \times 10^5 \text{ N/m}^2$. Estimate the molar mass of carbon dioxide from this data.

For an ideal gas pV/T is a constant for a particular mass of gas and if we consider one mole of the gas pV/T is a constant R_0. *Figure 3.18* shows how, for one mole of ammonia, the value of $pV/(R_0 T)$ depends on the pressure. For an ideal gas it should be independent of pressure.

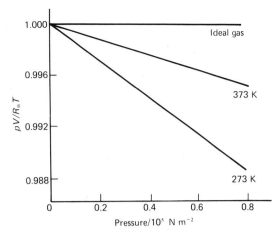

Figure 3.18 Variation of $pV/R_0 T$ with pressure for the molar mass of ammonia

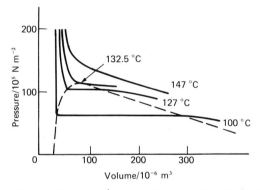

Figure 3.19 Pressure/volume relationship for a molar mass of ammonia

The higher the temperature the closer the ammonia comes to behaving like the ideal gas, though at any temperature it comes close to it when the pressure is low.

Figure 3.19 shows how the volume of one mole of ammonia varies with pressure (note that the pressures are much higher than in *Figure 3.18*). If the ammonia was an ideal gas and obeyed Boyle's law then the graph would look like *Figure 3.14*. At high temperatures and low pressures the graph does approximate to the ideal gas behaviour.

However, there is more happening to the ammonia than is immediately apparent from the graph. At temperatures below 132.5 °C the ammonia is not always a gas. At pressures and volumes in the area of the graph within the broken curve both liquid ammonia and ammonia vapour are present. The almost vertical graph lines, above the broken curve for temperatures below 132.5 °C, represent pressures and volumes where only liquid ammonia is present. Thus, it is when a substance is either liquid, or partially liquid, or close to these regions, that the ideal gas equation is not followed. This applies to all substances, even the so-called permanent gases.

The temperature of 132.5 °C for ammonia is known as the **critical temperature**. This is the highest temperature at which ammonia can be changed from a gas to a liquid. At temperatures above this the ammonia remains as a gas no matter what the pressure is. Below this temperature there is some pressure which will liquefy the ammonia.

The following are the critical temperatures for some substances.

Substance	*Critical temperature/* °C
Nitrogen	−147.1
Oxygen	−119.8
Carbon dioxide	31.1
Water	374.0

The so-called permanent gases nitrogen and oxygen have critical temperatures well below room temperature and experiments with these gases at pressures around atmospheric pressure thus show them as obeying the ideal gas laws. Carbon dioxide and water have critical temperatures which would imply that we should not expect them to obey the ideal gas laws exactly, unless the pressure was very low.

Suggestions for answers

1   The paint is known as thixotropic. There is no clear cut division by means of flow properties between solids and fluids.

2   The putty has the opposite behaviour to the paint.

3   $1 \text{ cm}^3 = 10^{-6} \text{ m}^3$, 0.1998 kg

4   Consider a room of height nominally 3 m, length 5 m, width 4 m. Volume = 60 m^3, hence mass is 72 kg.

5   0.810

6   7.800

7   $100/0.002 = 50000$ N/mm^2 or 5×10^{10} N/m^2

8   (a) Ah
(b) $Ah\rho$
(c) $Ah\rho g$
(d) $h\rho g$
(e) $100 \times 10^{-3} \times 1000 \times 9.8 = 980$ N/m^2

9   At the base where h in $h\rho g$ is the largest

10   $200 \times 10^{-3} \times 1000 \times 9.8 = 1960$ N/m^2. The gas pressure is higher than atmospheric pressure.

11   $760 \times 10^{-3} \times 13\,600 \times 9.8 = 101\,300$ N/m^2 to three significant figures

12   13.6 times greater, i.e., 10 300 mm or 10.3 m.

13   100 N

14   $10^{-6} \times 1000 \times 9.8 = 9.8 \times 10^{-3}$ N

15   $10^{-6} \times 1.2 \times 9.8 = 1.2 \times 10^{-5}$ N to two significant figures

16   Upthrust $= 0.008 \times 1.2 \times 9.8 = 0.094$ N.
Gravitational force $= 0.008 \times 0.09 \times 9.8 + 3 \times 10^{-3} \times 9.8$
$= 0.036$ N. Hence net upwards force is 0.058 N.

17   Gravitational force $= 917 \ Vg$, where V is total volume.
Upthrust $= V'g$, where V' is volume immersed. These must
be equal, hence $V'/V = 917/1000$. More than 90% of the cube is
below the surface. This calculation applies to icebergs, though
the sea water has a slightly higher density than 1000 kg/m^3.
Most of an iceberg is below the surface of the water. When you
consider the size of icebergs when viewed above the water this
makes them an even more terrifying size.

18   If the solid floats so that its entire volume V is below the surface
then $V\rho g = V\sigma g$, where ρ is the density of the solid and σ that
of the liquid. For this to happen $\rho = \sigma$. If the solid floats so that
only part of its volume V' is below the surface then $V\rho g = V'\sigma g$,
where V' is smaller than V. Thus σ must be greater than ρ. For
floating to occur the density of the solid must be equal to or less
than that of the liquid.

19   They have an average density lower than that of water, sufficiently
low to create enough upthrust to overcome the gravitational
force pulling the man down.

20   Gravitational force $= 70 \times g + 0.020 \times \rho \times g$, where ρ is the
density of the lifebelt. Upthrust $= 0.020 \times 1100 \times g + 0.8 \times V \times$
$1100 \times g$, where V is the volume of the man. Also $70 = 1200 \times V$.
For the upthrust to equal the gravitational force, the density of
the lifebelt must be 167 kg/m^3.

21   Ethyl alcohol. This is why some thermometers use alcohol. It
expands more than mercury and so gives a more sensitive thermometer.

22   171×10^{-6} °C^{-1}

23   Apparent coefficient $= 201 \times 10^{-6}$ °C^{-1}, hence change in
volume is 1.2 cm^3.

24   (a) $1 + 1 \times 10 \times 210 \times 10^{-6} = 1 + 2.1 \times 10^{-3} = 1.0021$ m^3.
(b) Decreased. Initially mass $= 1000 \times 1 = 1000$ kg. After
expansion density $= 1000/(1 \times 10 \times 210 \times 10^{-6} + 1) = 997.9$
kg/m^3.

25   The water near the bottom of the beaker will become warmer
than the water further up. Hot water has a lower density than
cold water and so rises through the cold water to the surface. The
water thus circulates within the beaker. Such movements are
called convection currents.

26   Twenty times greater

27   Air will enter the tyre when the pressure in the pump is 1.8×10^5 N/m^2. To compress the air from a pressure of 1.0×10^5 N/m^2 to 1.8×10^5 N/m^2 the volume must be reduced by the ratio 1.8/1.0.

28   (a) 293 K, (b) 253 K, (c) 473 K

29 $V/273 = 2V/T$, hence $T = 546$ K or 273 °C

30   (a) To obtain the greatest change in length per °C the cross-sectional area should be as small as possible. The larger the volume of the trapped air the greater the change in volume and hence the greater the change in length. Thus a long length of trapped air with a small cross-sectional area is required.
(b) The mercury is used purely as an index and thus should be as short as possible. A long thread will exert an additional pressure on the air and reduce its volume.
(c) If the mercury thread was very small there would be no noticeable difference in pressure if the tube were tilted. As, however, the thread may have a significant length, varying the angle of the tube by a large amount could change the pressure sufficiently to change the volume.
(d) A straight-line graph which if extrapolated back to the zero length point would reach this at −273 °C.

31   (a) $273 \times 50/312 = 43.75$ cm^3
(b) $373 \times 100/273 = 136.6$ cm^3

32 $350 \times 3 \times 10^5/300 = 3.5 \times 10^5$ N/m^2

33 $0.9p/291 = p/T$, hence T is 323 K, i.e., 50 °C

34 $\dfrac{10^5 \times V}{300} = \dfrac{25 \times 10^5 \times 0.1V}{T}$, hence, T is 750 K, i.e., 477 °C

35 $\dfrac{1.013 \times 10^5 \times 0.831}{293} = 287.3$ N m kg^{-1} K^{-1}

36   (a) $\dfrac{21 \times 10^5 \times 50 \times 10^{-3}}{290 \times 297}$ or about 1.22 kg.

(b) 0.87 kg

37   Assuming that the ideal gas equation applies, for 1 kg of CO_2 we have $10^5/(1.90 \times 283) = R$. This gives $R = 186$ N m kg^{-1} K^{-1}. Hence molar mass $= 8314/186$ or about 44.7 g. With the molar mass of carbon as 12 g and that of oxygen 16 g then the molar mass of CO_2 should be 44 g.

Further problems   No suggestions for answers are given for these problems.

38   What is the volume occupied by 5 kg of air under normal room conditions?

39   What is the relative density of wood which has a density of 750 kg/m^3?

40   A stone has a mass of 3 kg and rests on a table.
If the surface area of the stone in contact with the table is 100 cm^2 what is the pressure exerted on the table by the stone?

41   Estimate the pressure exerted on the floor by the leg of a chair with you sitting in the chair.

42   The water level in the cold water tank in a house is 2 m above the tap in the upstairs bathroom and 5 m above the tap in the downstairs bathroom.
What are the water pressures at each tap?

43   Girls wearing shoes with heels having a very small area of contact with the floor (stiletto heels) make a much bigger dent in the floor than those wearing shoes with much larger area heels.
Why? The force in each instance may be exactly the same.

44   A car has a mass of 2000 kg and is supported on four tyres. The air pressure in the tyres is initially 16 × 10^4 N/m^2 but then drops to 10 × 10^4 N/m^2. These pressures are gauge pressures and are the amounts by which the pressure in the tyre exceeds atmospheric pressure.
How can the same load be supported by these different pressures?
Estimate the contact area with the ground when the tyres are fully inflated.

45   A tube contains a vertical column of water. The column is 400 mm high and the tube has a cross-sectional area of 100 mm^2.
(a) What is the pressure due to the water at the base of the column?
(b) What is the force acting on the closed base of the tube?

46   In 1648 B. Pascal found that the mercury in a barometer tube dropped in level by about 75 mm when the barometer was carried from the foot of the Le Puy de Dome mountain to its top. Air has a density of about 10^{-4} of that of mercury.
Estimate the height of the mountain above its base.

47   If you get a tube full of liquid, perhaps a drinking straw up which you have sucked liquid, and then put your thumb over the upper end the liquid will not flow out of the open lower end as long as you keep your thumb over the upper end. When you remove your thumb the liquid runs out.
Explain.

48   A piece of wood is 600 mm long, 200 mm wide and 50 mm thick. It has a density of 700 kg/m^3.
What part of the thickness of the wood will be below the surface when it floats in water of density 1000 kg/m^3?

49   A Coca-Cola can has had some sand placed in it so that it floats vertically in water of density 1000 kg/m^3 with half of its 120 mm height immersed. The can has a diameter of 65 mm.
How much of the can will be below the liquid surface if it is then put in sea water of density 1100 kg/m^3?

50   A solid block of wood will float, a solid block of metal will not float in water.
Why then can ships be made of metal rather than wood?

51   Explain the principle of the hot air balloon.

52   The internal volume of a glass flask, up to a reference mark on the stem, is exactly 1000 cm^3 at 20 °C. The flask is filled with water up to this mark.
What will be the volume of water above the mark when the temperature is 40 °C. The coefficient of expansion of the water is 210×10^{-6} °C^{-1} and that of the glass 9×10^{-6} °C^{-1}.

53   A method that has been used for the determination of the coefficient of expansion of a liquid is to weigh an object first in air and then submerged in the liquid at two different temperatures.
Explain how these readings could be used to give the coefficient of expansion.

54   (a) Explain the convection currents that would arise in a pond when the upper surface cools from, say, 20 °C to 15 °C. Assume that initially all the water in the pond was at the same temperature.
(b) Explain the convection currents that would arise in a pond when the upper surface cools from 4 °C to 0 °C.

55   A bubble of air rises from the bottom of a lake 10 m deep.
How will the volume of the bubble at the surface be related to its volume at the bottom of the lake if there is no change in temperature?

56   A gas having a volume of 0.4 m^3 under a pressure of 2×10^5 N/m^2 is to be compressed into a volume of 0.1 m^3, what pressure will be required?

57   What are the temperatures on the Kelvin scale of (a) 18 °C, (b) 400 °C, (c) −50 °C?

58   What are the temperatures on the Celsius scale of (a) 50 K, (b) 500 K, (c) 1000 K?

59 A balloon is partly inflated using helium at a temperature of 15 °C and assumes a volume of 120 m^3.
What will be the volume of the balloon if the temperature rises to 28 °C? Assume the pressure remains constant.

60 A car tyre contains air at a pressure of 2.5 × 10^5 N/m^2.
Guess what might happen to the pressure if the car stands in the sun and the temperature of the tyres increases.

61 At the bottom of a lake the temperature is 7 °C and the pressure 3 × 10^5 N/m^2. At the surface the temperature is 17 °C and the pressure 1 × 10^5 N/m^2.
How would the volume of an air bubble produced at the lake bed change as it moved to the surface?

62 Gas is stored in a cylinder at a pressure of 4.0 × 10^5 N/m^2 and a temperature of 17 °C.
What will be the pressure at 27 °C if the container does not change in volume and no gas escapes?

63 A car tyre contains air at a pressure of 3 × 10^5 N/m^2, the air occupying a volume of 6 × 10^{-3} m^3. The temperature is 27 °C.
Estimate the mass of air in the tyre.

64 Estimate the mean molecular mass of air from the following data. A glass bulb weighs 24.2351 g when evacuated and 24.8146 g when full of air at a pressure of 1.01 × 10^5 N/m^2 and a temperature of 27.0 °C. The bulb has a volume of 5.00 × 10^{-4} m^3.

65 The following data are for argon at a constant temperature of −50 °C.
Does argon behave like an ideal gas at this temperature?

Pressure/ 10^5 N m^{-2}	8.99	17.65	26.01	34.10	41.92	49.50	56.86
Volume/ 10^{-3} m^3	2.00	1.000	0.667	0.500	0.400	0.333	0.286

4 Fluid flow

Concorde

Objectives

The intention of this chapter is to show how a simple consideration of fluid flow enables you to explain many everyday phenomena. You will need an understanding of Newton's laws.

The general objectives for this chapter are that after working through it you should be able to:

(a) Define the term streamline and describe streamline patterns round a sphere and a wing section;
(b) Derive and use the equation of continuity;
(c) Argue the case for a pressure drop occurring where streamlines come closer together;
(d) Explain how an aircraft obtains lift, and other similar situations involving a pressure drop;
(e) Derive Bernoulli's equation and use it in problems;
(f) Explain what is meant by the term boundary layer;
(g) State Newton's law of viscosity and use it in problems;
(h) Describe the streamline patterns occurring with different velocity flow past a sphere and use the pattern to describe some everyday situations;
(i) Explain the significance of Reynolds' number.

Teaching note

Ideas for experiments and background reading appropriate to this chapter can be found in *Shape and Flow* by A.H. Shapiro.

Flow patterns

The aircraft in the opening picture is moving through air, a fluid.
A car moving along a road is moving through air, a fluid. Ships move
through water, a fluid. There are many instances where objects move
through fluids or fluids flow past objects. How do fluids flow past
objects?

The way in which a fluid flows past an object can be made visible
by introducing some form of 'float' and observing how this moves.
For liquids, gas bubbles and drops of dye have been used as floats.
For gases, smoke and balloons have been used.

Figure 4.1 shows some flow patterns. *Figure 4.1(a)* is for flow past
a cylinder, an unstreamlined object. *Figure 4.1(b)* is for flow past a
streamlined object. The difference between the two flow patterns
which justifies one being called **streamlined** and the other **unstreamlined** is that the flow in the wake of the unstreamlined object is in the
form of eddies which follow no particular pattern while that behind
the streamlined object shows little disturbance.

*Figure 4.1 Flow in the wake of (a) an
unstreamlined object, (b) a streamlined
object*

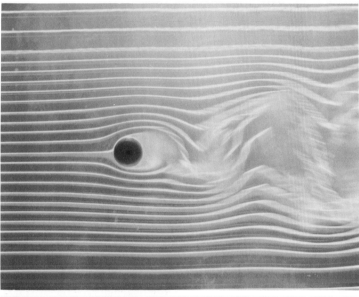

(a)

(b)

The lines shown by the movement of the markers are called streamlines. A **streamline** is defined as a line, the tangent to which at any point gives the direction of fluid motion at that point. To find the direction of flow at any point along a streamline all that you have to do is draw the tangent to that streamline at the point concerned (*Figure 4.2*). As the flow is in a direction tangential to a streamline there is no fluid velocity at right angles to a streamline. As will be evident from the figures, the shapes of the streamlines are determined by the objects the fluid is moving past.

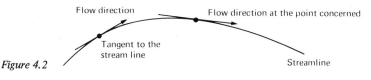

Figure 4.2

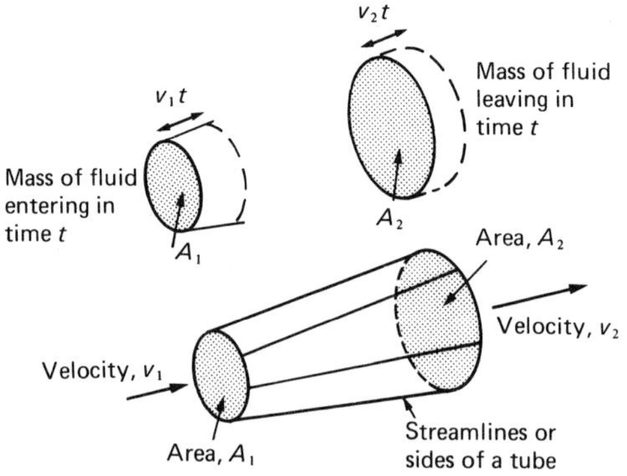

Figure 4.3

Consider the flow of a fluid down a tube formed by a set of streamlines — it could be an actual tube with the streamlines hugging the surface of the tube (*Figure 4.3*). There is no flow across the streamlines, or tube walls, thus the mass of fluid entering the tube in time t must equal the mass of fluid leaving the tube in time t. If v_1 is the velocity with which the fluid enters the tube and if A_1 is the cross-sectional area of the inlet, then in time t a volume $A_1 v_1 t$ will enter the tube. The mass entering is thus $\rho_1 A_1 v_1 t$, where ρ_1 is the density of the fluid entering. Similarly the mass leaving in time t will be $\rho_2 A_2 v_2 t$, where ρ_2 is the density of the fluid leaving with a velocity v_2 through an area A_2. As the mass entering in time t equals the mass leaving in the same time

$$\rho_1 A_1 v_1 t = \rho_2 A_2 v_2 t$$

Hence

$$\rho_1 A_1 v_1 = \rho_2 A_2 v_2$$

This equation is known as the **equation of continuity**.

If the fluid leaves with the same density as it enters, i.e., the fluid is considered to be incompressible, then

$$A_1 v_1 = A_2 v_2$$

This is generally the case for liquids. In some cases for gases the density does change.

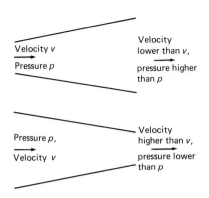

Figure 4.4 (a) Streamlines moving further apart, (b) streamlines moving closer together

For an incompressible fluid an increase in the area through which the fluid flows must mean a drop in fluid velocity. Thus, in the stream-line patterns shown in *Figure 4.1(a)* and *(b)*, where the streamlines move further apart the velocity must decrease; where the streamlines move closer together the velocity increases.

When velocity increases there is an acceleration. To produce an acceleration requires a net force (Newton's second law of motion). Thus, when the velocity of a fluid increases there must be a force producing the acceleration. In a fluid the net force in the direction of the accelera-tion is produced by a pressure difference. When a fluid accelerates it is going from one pressure to a lower pressure. If the fluid is decelerating, i.e., moving from one velocity to a lower velocity, then the fluid is moving from one pressure to a higher pressure (*Figure 4.4*).

Questions   1   *Figure 4.5* shows a section of a fire-hose nozzle. How, in general, does the velocity with which the water emerges from the nozzle compare with the velocity with which it enters the nozzle?

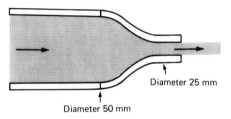

Figure 4.5 A fire-hose nozzle

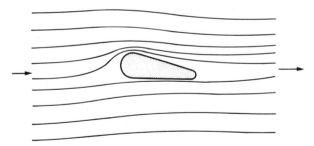

Figure 4.6 Streamlines past an aircraft wing section

2   *Figure 4.6* shows the streamlines for air flow past a section of the wing of an aircraft. How does the velocity of the air change as it moves over the wing when compared with its movement under the wing? How will the pressure above the wing compare with the pressure below the wing? What effect will this pressure difference have on the wing?

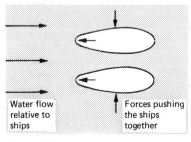

Figure 4.7

3   When two ships move on closely parallel courses (*Figure 4.7*) care has to be exercised to prevent the ships being pulled together and so colliding. What is the origin of the force which tends to pull the ships together?

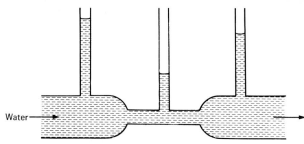

Figure 4.8

4   *Figure 4.8* shows a piece of apparatus you may come across in the laboratory. Why does the vertical tube in the central section show a lower level of water than the two outer vertical tubes?

5   *Figure 4.9* shows a paint spray. Why is the paint sucked up the tube and sprayed out through the nozzle?

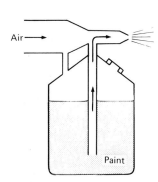

Figure 4.9 A paint spray

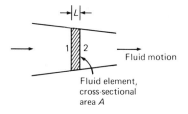

Figure 4.10 Fluid moving horizontally

Bernoulli's equation

The following is a derivation of a relationship known as **Bernoulli's equation** by the application of Newton's second law of motion to a fluid. An alternative derivation can be obtained from a consideration of the conservation of energy for the moving fluid.

Initially we will consider the simple case of a fluid moving with streamline motion in a horizontal direction. The fluid is accelerating as a result of moving from a high pressure area to a low pressure area. We consider the motion of a small element of this fluid (*Figure 4.10*). The element is the block of fluid between 1 and 2 in the figure. If the pressure at position 1 is p_1 then the force acting on that end of the element is p_1A, where A is the cross-sectional area of the element. If the pressure at the other end of the element, position 2, is p_2 then the force acting on that end of the element is p_2A. The net force acting on the element due to this difference in pressure is thus

$$\text{Net force} = p_1A - p_2A = (p_1 - p_2)A$$

There are no other forces acting on the element in its direction of motion and so this net force must be equal to ma, where m is the mass of the element and a its acceleration.

$$(p_1 - p_2)A = ma$$

The volume of the fluid element is LA and hence if it has a density ρ the mass m is $LA\rho$. Hence

$$(p_1 - p_2)A = LA\rho a$$

The acceleration a means that there is a change in velocity occurring. An equation relating the initial and final velocities is $v^2 = u^2 + 2as$ (see *Book 1: Motion and Force*, Chapter 1). Thus if the velocity is v_1 at position 1 and v_2 at position 2 then as these two points are a distance L apart

$$v_2^2 = v_1^2 + 2aL$$

Thus

$$\frac{v_2^2 - v_1^2}{2} = aL$$

and so

$$(p_1 - p_2)A = \frac{A\rho(v_2^2 - v_1^2)}{2}$$

This equation can be rearranged to give

$$p_1 + \rho\frac{v_1^2}{2} = p_2 + \rho\frac{v_2^2}{2}$$

This is the form of Bernoulli's equation for fluid flow along the horizontal direction, the fluid neither gaining nor losing height during the flow. The equation tells us that in any such case of streamline flow the sum of the pressure at a cross section of the tube and the product of the density and half the square of the velocity at that point is equal to the same sum at any other point in the same fluid flow.

$$p + \frac{\rho v^2}{2} = \text{a constant}$$

For the above expression to remain constant an increase in pressure must mean a decrease in velocity and conversely a decrease in pressure must mean an increase in velocity.

Question   6   For water flowing through the apparatus shown in *Figure 4.8* the pressure is found to drop by 1000 N/m² on flowing from the wide tube to the narrow tube.
How are the velocities in the wide and narrow tubes related? If the wider tube has a diameter of 12 mm and the narrow tube 6 mm what is the volume of water flowing through the apparatus per second?

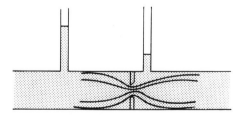

Figure 4.11 The orifice plate flowmeter

The apparatus shown in *Figure 4.8* is called a **Venturi tube**. By measuring the pressure difference that occurs when a fluid passes from a wide tube to a narrower tube the velocities of the fluid can be calculated and hence the volume flowing through the tube per second.
The Venturi tube is an instrument for the measurement of flow rate.
    A simpler form of the Venturi tube is the **orifice plate** (*Figure 4.11*). This consists of an orifice which is inserted into the pipe to give the

change in the size of the flow tube. The pressure difference is measured between a point some distance in front of the orifice and a point immediately behind the orifice. The theory is the same as for the Venturi tube.

In the discussion of the Venturi tube and the orifice plate no allowance has been made for frictional effects. The velocity of the fluid as it flows along a tube or through an orifice will fall below the theoretical value because of friction. Corrections have to be made for this when such instruments are used for flow-rate measurements.

Question   7   A Venturi tube is used to measure the flow rate of water through a horizontal pipe. If the wide tube has a diameter of 300 mm and the constriction a diameter of 150 mm what will be the flow rate when the pressure difference indicated is 1100 N/m². Neglect any frictional effects.

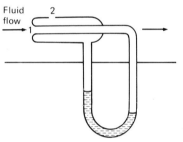

Figure 4.12 A version of the Pitot tube

The airspeed indicator of an aircraft is used to determine the velocity of the aircraft relative to the surrounding air. An alternative way of considering it is as a measurement of the speed of the air moving past the aircraft. The basis of this instrument is called a **Pitot tube** (*Figure 4.12*). The fluid entering the front opening of the Pitot tube comes to rest and, thus, for position 1 the velocity is zero. The opening at 2 has the fluid moving past it with the fluid velocity v. The pressure difference between points 1 and 2 is indicated by the manometer. Thus, applying Bernoulli's equation

$$p_1 + 0 = p_2 + \rho \frac{v^2}{2}$$

$$\text{Pressure difference} = p_1 - p_2 = \rho \frac{v^2}{2}$$

The pressure difference is thus proportional to the square of the fluid velocity.

Question   8   A Pitot tube of the form shown in *Figure 4.12* is used to measure air speed. A mercury manometer shows a pressure difference of 12 mm. What is the air speed?

The Bernoulli equation so far considered only applies to situations where the flow is in the horizontal plane and no gain or loss of height occurs. Consider a situation, as in *Figure 4.13*, where, in addition to the cross section of the flow tube varying, the fluid gains height. As before, the net force acting on the element due to the difference in pressure between positions 1 and 2 is

$$\text{Net force} = (p_1 - p_2)A$$

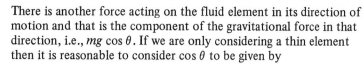

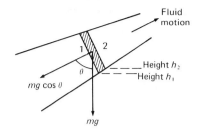

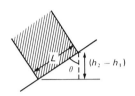

Figure 4.13 Fluid moving upwards

There is another force acting on the fluid element in its direction of motion and that is the component of the gravitational force in that direction, i.e., $mg \cos \theta$. If we are only considering a thin element then it is reasonable to consider $\cos \theta$ to be given by

$$\cos \theta = \frac{(h_2 - h_1)}{L}$$

Thus the gravitational force component is given by

$$\frac{mg(h_2 - h_1)}{L}$$

But the mass of the element m is ρAL, hence the gravitational force component is

$$\rho Ag(h_2 - h_1)$$

Therefore the net force acting on the element and causing it to accelerate is

$$(p_1 - p_2)A - \rho Ag(h_2 - h_1)$$

This must be equal to ma. As before we can write m in terms of A, ρ and L and a in terms of v_1, v_2 and L. Hence

$$(p_1 - p_2)A - \rho Ag(h_2 - h_1) = \frac{A\rho(v_2^2 - v_1^2)}{2}$$

This equation can be rearranged to give

$$p_1 + \rho \frac{v_1^2}{2} + \rho gh_1 = p_2 + \rho \frac{v_2^2}{2} + \rho gh_2$$

This is the full form of **Bernoulli's equation**. The equation can only be applied when the motion is streamline with no turbulence and to incompressible fluids, i.e., fluids in which density can be considered to be independent of pressure over the pressures involved. There is also another factor — we have ignored any frictional effects. A fluid which flows without any frictional effects is an 'unreal' fluid and the discussions we have so far advanced have been referred to as 'the flow of dry water'.

Figure 4.14 shows a simple situation, the flow of a fluid out of an orifice in the side of a container. If p_1 is the pressure on the surface of the fluid in the container and v_1 the velocity of that surface, p_2 the pressure at the efflux point 2 and v_2 the fluid velocity at that point then

$$p_1 + \rho \frac{v_1^2}{2} + \rho gh = p_2 + \rho \frac{v_2^2}{2}$$

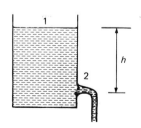

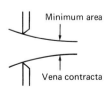

Figure 4.14

If the container is open to the atmospheric pressure and the efflux is into the atmosphere, p_1 and p_2 are virtually identical. If the surface area of the fluid surface in the container is considerably larger than the area of the cross-section of the fluid emerging from the orifice then v_1 is considerably smaller than v_2 and so v_1^2 can be neglected in comparison with v_2^2. The equation can thus be simplified to

$$\rho gh = \rho \frac{v_2^2}{2}$$

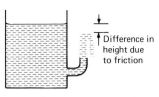

Difference in height due to friction

Figure 4.15

Hence

$$v_2^2 = 2gh$$

This is the same velocity as that acquired by any body falling through a height h. This relationship is known as **Torricelli's theorem.**

As the fluid passes through the orifice it is found that its minimum cross-sectional area is smaller than the area of the orifice. This minimum area occurs some distance from the orifice and is known as the **vena contracta**. This is effectively the area of the orifice.

There is also some friction, as can be shown by comparing the height from which the fluid falls and the height to which it will go if directed upwards on emerging from the orifice (*Figure 4.15*). If there was no friction the emerging stream should reach the same height as the fluid surface in the container.

Question   9   A tank of water has a small hole in its side at a point 1.2 m below the water surface. If the tank is open to the atmosphere and the water emerges from the hole into the atmosphere what is the velocity of the emerging water, if friction is neglected? If the vena contracta of the jet of escaping water has an area of 4 mm² what is the volume of water escaping from the tank per second?

Fluid flow past surfaces

If you have ever tried to pour treacle out of a tin you will have realized that it 'clings' to the sides of the can and shows a marked reluctance to slide out. Even when you have poured out the treacle you will invariably find a layer still clinging to the surface. Liquids and gases tend to cling to surfaces when you try to move them relative to the surface. Pouring the treacle out of the tin, fluid moving relative to a surface, or moving a spoon through the treacle, a surface moving relative to the fluid, present the same problems.

If you put a pack of playing cards on a table (*Figure 4.16*) and push with a sliding action against the top card you will get the cards to 'flow'. The result of the push is to move each card parallel to its neighbour — and the higher the card is up the pack the greater the distance it will have moved from its original position. A clinging action occurs between the bottom card and the table. This clinging action you would account for by talk of frictional forces between the bottom card and the surface. There is also a frictional force between each card and the next. The slope of the pack is larger the larger the push sliding force you apply.

If you blow across the top of a dusty table you will find that not all the dust blows off. The smaller particles of dust, those very close

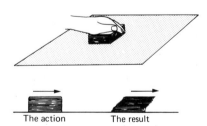

The action   The result

Figure 4.16 The force applied by the hand to the cards is called a shear force

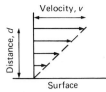 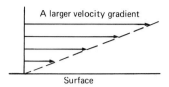

Figure 4.17 The arrow-headed lines represent the velocities, the longer the line the greater the velocity. Velocity gradient is v/d

to the surface, cling to the surface and no matter how hard you blow they do not move. It appears that the air immediately adjacent to the surface has no velocity, although the air further above the surface has a velocity. The air flow past the surface has produced a velocity gradient (*Figure 4.17*).

When there is relative motion between a fluid and a surface the fluid immediately adjacent to the surface has the same velocity as the surface — it clings to the surface.

Question 10 Why can you not pour all the water out of a beaker?

When a fluid is moving past a surface the velocity of the fluid will be changed in the region close to the surface. This region is called the **boundary layer.** The thickness of this boundary layer depends on the fluid concerned and the velocity of the fluid. For air moving over the land surface at about 10 m/s the boundary layer is about 100 m thick. For water at room temperature flowing past a plate 0.1 m long with a velocity of about 10 m/s the boundary layer is about 5 mm thick.

Within the boundary layer there is a velocity gradient. The velocity gradient occurs because each successive layer of fluid exerts a drag on the next layer — like the playing cards. This discussion relates to what is called **laminar flow** and assumes there is no turbulence. The zero velocity layer adjacent to the surface exerts a retarding force on the layer immediately above it and slows it down to below the full fluid velocity. This layer exerts a retarding force on the layer immediately above it, and so on until, at the edge of the boundary layer, the retarding effects become insignificant and the velocity of the layer is the full fluid velocity. The presence of the surface thus exerts a drag on the fluid. This force is rather like a frictional force between two solids — opposing relative motion. The opposing force on a fluid moving relative to a surface is called a **viscous force** and the fluid is said to show **viscosity.** Viscosity may be thought of as the internal friction of a fluid. Because of viscosity a force must be exerted to move a surface through a fluid, pour a liquid out of a container, or pass a fluid through a pipe.

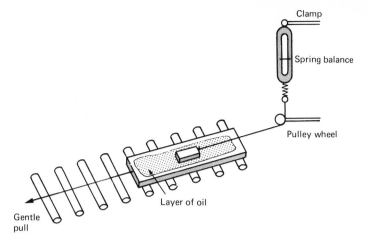

Figure 4.18 An experiment to investigate viscous forces. (Compare with the frictional force experiment in Book 1: Motion and Force, *Chapter 2)*

Figure 4.18 illustrates a simple experiment in which one plate is moving past another with a layer of liquid between them. When a force is applied to one of the plates it moves, after an initial acceleration, with a reasonably constant velocity for a constant force. The viscous drag force depends on the plate area, the velocity and the thickness and type of the liquid between the plates. For many fluids the relationship between these quantities is found to be

$$\frac{\text{Force}}{\text{Area}} \propto \frac{\text{velocity}}{\text{distance between plates}}$$

The velocity divided by the distance between the plates, i.e., liquid film thickness, is the velocity gradient.

$$\frac{\text{Force}}{\text{Area}} \propto \text{velocity gradient}$$

$$\frac{\text{Force}}{\text{Area}} = \text{coefficient of viscosity} \times \text{velocity gradient}$$

The **coefficient of viscosity** (η) depends on the fluid concerned.

Units: force — N, area — m^2, velocity — m/s, distance — m, velocity gradient — (m/s)/m or s^{-1}, η — N s m^{-2} or Pa s.

Fluids which obey the above relationship are called **Newtonian fluids** since the relationship was first proposed by Newton. For liquids the coefficient of viscosity generally decreases quite noticeably with temperature, for gases there is an increase with temperature but this is not so noticeable. The following are some typical coefficients of viscosity.

Fluid	Temperature/°C	Coefficient of viscosity/ N s m^{-2}
An oil	0	5.3
	20	0.99
	40	0.23
Water	0	0.017 9
	20	0.010 0
	40	0.006 5

Fluid	Temperature/°C	Coefficient of viscosity/ $\mathrm{N\,s\,m^{-2}}$
Air	0	0.000 017
	20	0.000 018
	40	0.000 019

A large coefficient of viscosity means that a large force is needed to slide one plate over another when that fluid is between the plates. It also means that such a fluid flows less readily out of a bottle or through a pipe, or past a surface.

Questions   **11**   Would you consider treacle to have a higher or a lower coefficient of viscosity than water? State the evidence on which you base your answer.

12   Water flows over a horizontal plate 2 m long and 1 m wide. At a distance of 10 mm from the plate the water velocity is 5 m/s. What is the size and direction of the horizontal force on the plate due to the moving water? Assume that out to this distance of 10 mm from the plate the water velocity increases in direct proportion to the distance. Take the coefficient of viscosity to be $0.010\ \mathrm{N\,s\,m^{-2}}$.

13   Why is a pressure difference between the two ends of a tube needed to keep the fluid moving with constant velocity along the pipe? This question is similar to that of why is a force needed to keep an object moving with a constant velocity?

14   The volume of fluid flowing through a pipe per second $\dot{V}$ is related to the pressure difference p between the ends of the pipe by

$$\dot{V} = \frac{\pi p r^4}{8\eta L}$$

where L is the length of the pipe, the distance across which the pressure difference occurs, r the pipe radius and η the coefficient of viscosity of the fluid. This is known as **Poiseuille's equation** and applies to non-turbulent flow where the streamlines are parallel to the sides of the tube.

(a) What is the rate of flow of water through a pipe of radius 2.5 mm and length 1 m if the pressure difference between the two pipe ends is $100\ \mathrm{kN/m^2}$. Take the coefficient of viscosity to be $0.010\ \mathrm{N\,s\,m^{-2}}$.
(b) Do you need a larger or smaller pressure difference for the same rate of flow of a more viscous fluid?

15   A grease nipple on a car has an internal diameter of 0.5 mm and a length of 2 mm. What pressure difference is needed to force $0.6\ \mathrm{cm^3}$ of grease through the nipple in 1 s? What would the pressure difference be if it took 10 s to force the same volume through the nipple? Take the coefficient of viscosity to be $50\ \mathrm{N\,s\,m^{-2}}$ and assume that Poiseuille's equation holds.

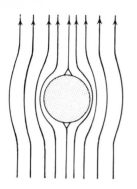

Figure 4.19 Streamlines past a sphere at low velocities

If you drop a ball-bearing in a vacuum it moves with a constant acceleration for its entire fall. If you drop the same ball-bearing in oil, although initially it accelerates the acceleration is not constant and gradually diminishes until the ball is moving with a constant velocity (known as the **terminal velocity**). For a sphere moving at low velocities such that there is no turbulence at all, i.e., the streamlines are as shown in *Figure 4.19*, the viscous drag force acting on it is $6\pi\eta rv$, where r is its radius and v the velocity at any instant. This is known as **Stokes' law.** The forces acting on the sphere during its fall are thus:

Gravitational force, downwards $= mg$

where m is the mass of the sphere and g the acceleration due to gravity.

Upthrust $= V\sigma g$

where σ is the fluid density, V the volume of the sphere, and

Viscous drag force, upwards $= 6\pi\eta rv$

The viscous drag force increases as the sphere accelerates. At some particular velocity the drag force becomes large enough to result in zero net force acting on the sphere. When this happens the sphere moves with a constant velocity. Then

$$mg = V\sigma g + 6\pi\eta rv$$

As $V = 4\pi r^3/3$ and $m = V\rho$, where ρ is the density of the sphere, the equation can be simplified to

$$v = \frac{2r^2(\rho - \sigma)g}{9\eta}$$

Question 16   Estimate the terminal velocity reached by a tiny raindrop of radius 0.05 mm falling in still air if Stokes' law is assumed to hold.

Stokes' law, the viscous drag being proportional to the velocity, applies when there is no, or very little, turbulence produced by the sphere moving through the fluid. At higher velocities turbulence occurs and the viscous drag becomes proportional, roughly to the square of the velocity.

Figure 4.20 shows how the flow pattern changes as the velocity of the sphere through the fluid increases. At first (*Figure 4.20(a)*) eddies (sometimes called vortices) are formed immediately behind the sphere. Then at higher velocities (*Figure 4.20(b)*) these break away and travel with the flow. The eddies peel off alternately from one side and then the other — leading to an oscillating disturbance behind the sphere. At higher velocities the eddies form a turbulent band behind the sphere. This oscillating disturbance has many applications, organ pipes and reed instruments use the oscillation to produce a musical

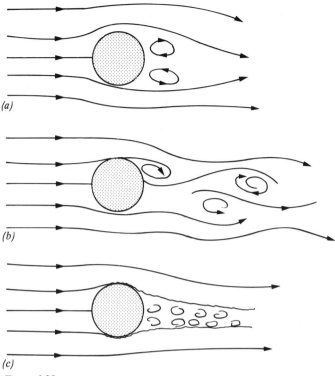

(a)

(b)

(c)

Figure 4.20

note, and consequences. Flags flutter and dangerous oscillations of buildings can occur (see *Book 6: Rays, Waves and Oscillations*). At yet higher velocities (*Figure 4.20(c)*) there is a marked reduction in the drag force when the fluid in the boundary layer starts to become turbulent. The velocity at which this occurs depends on the roughness of the sphere surface: for a rough surface the reduced drag force occurs at a lower velocity than for a smooth one. The drag force on the average pass of the ball in a soccer game is proportional to v^2, a well-hit goal kick is likely to have the reduced drag and so carry much further than might otherwise be expected.

Question   17   Explain why flags flutter.

Reynolds' number

In the 19th century Osborne Reynolds investigated the flow of fluids through pipes with a view to finding the conditions for which the flow would be streamline and those for which it would be turbulent. He injected dye into the centre of tubes through which fluids were flowing and observed whether the dye followed a straight line path down the centre of the tube, i.e., streamline flow, or a disturbed path, i.e.

turbulent flow. After investigating many variables he obtained the following relationship

$$R_n = \frac{\rho v d}{\eta}$$

where ρ is the fluid density, η its coefficient of viscosity, d the tube diameter and R_n a number called **Reynolds' number.** If the number is below about 2000 the flow is streamline, if more than about 3000 turbulent. For numbers between 2000 and 3000 the flow is unstable and could be either streamline or turbulent.

Question   18   Would you expect streamline or turbulent flow for water flowing at 0.05 m/s along a pipe of diameter 20 mm? Take the density of water to be 1000 kg/m^3 and the coefficient of viscosity as 0.01 N s m^{-2}.

Similar expressions can be obtained for shapes other than tubes and values of Reynolds' number determined for which the flow changes from streamline to turbulent.

Reynolds' number has another use. To investigate the flow of a fluid over new shapes of aircraft or other objects it is usual to carry out experiments with models before producing the full-size real object. Thus a scaled-down model of an aircraft might be tried in a wind tunnel before the full-scale aircraft is built. If the flow past the model has the same Reynolds' number as the flow past the full-size object then the flows will behave in the same way. The two situations are said to be **dynamically similar.**

Question   19   The flow of oil with a mean velocity of 0.4 m/s through a pipe of diameter 1 m is to be modelled by water flowing through a pipe of diameter 200 mm. What should be the mean velocity of the water flow? Take the coefficient of viscosity of the oil to be 1 N s m^{-2} and its density 900 kg/m^3.

Suggestions for answers   1   Velocity in the nozzle four times greater than in the pipe, this being the inverse ratio of the areas.

2   The pressure is lower above the wing than below the wing, hence the wing experiences lift. This pressure difference is brought about by the air flowing faster over the top of the wing than underneath it.

3   The higher velocity flow between the ships results in a pressure drop between the ships and hence a sideways force on each ship.

4   The higher velocity flow in the narrower tube produces a pressure drop.

5   The higher velocity air flow over the top of the vertical tube produces a pressure drop.

6   Pressure drop $= \rho(v_2^2 - v_1^2)/2$, hence $2 = v_2^2 - v_1^2$. $A_1 v_1 = A_2 v_2$, hence $d_1^2 v_1 = d_2^2 v_2$ and so $4v_1 = v_2$. Thus v_1 is about 0.36 m/s and v_2 about 1.46 m/s. The volume per second $= A_1 v_1 = A_2 v_2$, hence volume per second is about 40.7×10^{-6} m^3/s.

7   Pressure drop $= \rho(v_2^2 - v_1^2)/2$, hence $2.2 = v_2^2 - v_1^2$. $A_1 v_1 = A_2 v_2$, hence $d_1^2 v_1 = d_2^2 v_2$ and so $4v_1 = v_2$. Thus v_1 is about 0.38 m/s. The volume per second $= A_1 v_1$, hence the flow rate is about 2.71×10^{-2} m^3/s.

8   Pressure difference $= h\rho g = 12 \times 10^{-3} \times 13\,600 \times 9.8$ N/m^2. Hence $v^2 = 12 \times 10^{-3} \times 13\,600 \times 9.8 \times 2/1.3$; the velocity is about 49.6 m/s.

9      $v^2 = 2 \times 9.8 \times 1.2$, hence v is about 4.9 m/s. Volume per second $= Av$ or about 1.96×10^{-7} m^3/s.

10   Some 'sticks' to the beaker walls.

11   Treacle has a higher coefficient of viscosity because a larger force is needed to produce a given velocity gradient, e.g., to get it to flow out of a bottle.

12      $F = 0.010 \times 2 \times 1 \times 5/0.010 = 10$ N. The direction of the force is in the direction of the flow. The force needed to move the fluid has an opposite and equal force acting on it.

13   Because there are viscous forces. In the absence of such forces the fluid once set in motion would continue in motion with a constant velocity (Newton's first law).

14   (a) $\dot{V} = \dfrac{\pi \times 100 \times 10^3 \times (2.5 \times 10^{-3})^4}{8 \times 0.010 \times 1}$

        or about 15×10^{-5} m^3/s

        (b)   A larger pressure difference

15   About 7.8 MN/m^2, 0.78 MN/m^2

16      $\dfrac{2 \times 0.000\,05^2 \times (1000 - 1.2) \times 9.8}{9 \times 0.000\,018}$ or about 0.30 m/s

17   The leading edge of the flag produces eddies, as in *Figure 4.20*, which travel down the flag.

18 $R_n = \dfrac{1000 \times 0.05 \times 0.020}{0.01} = 100$

The motion would be expected to be streamline

19 $\dfrac{900 \times 0.4 \times 1}{1} = \dfrac{1000 \times v \times 0.200}{0.01}$

Hence v needs to be 0.018 m/s

Further problems   No suggestions for answers are given for these problems.

20   A liquid flows through a pipe such that where the pipe has a diameter of 10 mm the velocity is 30 m/s.
What will be the diameter of the tube at a point where the fluid velocity has dropped to 10 m/s?

21   A fire-hose nozzle has an inside exit diameter of 20 mm, the hose prior to this having a diameter of 60 mm.
If the water leaves the nozzle with a velocity of 70 m/s what is the water velocity in the hose and the volume of water emerging per second?

22   What pattern of streamlines might you expect to find round a fish swimming in the sea?

23   What effects might you expect when the wind blows through some of the narrow streets, roads and alleys of a city?

24   Look at the water streaming out of a tap. When a water column is moving vertically downwards the cross-sectional area of the stream changes.
Why?

25   Explain how the Venturi tube can be used for the measurement of flow rates.

26   Water flows along a horizontal pipe which tapers from 160 mm diameter to 120 mm diameter.
If the rate of flow along the pipe is 0.4 m^3/minute what will be the pressure difference between the two ends of the pipe?

27   If the pipe in question **26** is tilted so that the water flows down the pipe, and the difference in height between the two ends is 1.2 m, what will be the pressure difference between the two ends for the same rate of flow?

28   An open water tank discharges water to the atmosphere through a pipe.
If this discharge pipe has a diameter of 60 mm and the water leaves the pipe at a vertical height of 4 m below the water surface

in the tank, what will be the velocity with which the water leaves the pipe and the volume discharged per second?

29 An oil pipeline tapers from a diameter of 120 mm to 80 mm. If the pipe is horizontal and carries oil of density 850 kg/m^3 what will be the flow rate if the pressure difference between the wide and narrow parts of the tube is 30 kN/m^2?

30 A Pitot tube used for the measurement of the speed of an aircraft uses water as the manometric liquid. The difference in levels of the water in the U-tube manometer is 50 mm. What is the air speed if the air has a density of 1.3 kg/m^3?

31 In an experiment to determine the coefficient of viscosity of water it was found that a pressure difference due to a pressure head of 56 mm of water caused water to emerge from the open end of a capillary tube of internal radius 0.50 mm and length 100 mm at the rate of 8.1 cm^3 per minute. Calculate the coefficient of viscosity if the motion is assumed to be streamline.

32 A steel ball bearing of density 7700 kg/m^3 and diameter 3 mm falls through oil of density 1300 kg/m^3. After an initial acceleration the ball is found to move with a constant velocity such that it covers 250 mm in 7.5 s. Calculate the coefficient of viscosity of the oil.

33 Two raindrops of the same size fall through still air with a velocity of 120 mm/s. If the drops collide and coalesce what might you expect the new velocity to be? What assumptions do you need to make to arrive at your answer.

34 A powder which is insoluble in water is shaken with water and allowed to stand. The depth of the water is 120 mm. What is the largest diameter particle that will remain in suspension one hour later? The density of the particles is 2800 kg/m^3 and all the particles may be assumed to be spherical.

35 How would the result in question 34 differ if the suspension was placed in a centrifuge which gave an effective acceleration due to 'gravity' of ten times that due to the earth's gravity?

36 Describe an experiment to measure, or compare with a standard, the coefficient of viscosity of water.

37 Water with a viscosity of 0.01 N s m^{-2} flows with a velocity of 500 mm/s through a pipe of diameter 4 mm. What is the Reynolds' number and what would you expect the nature of the flow to be?

38 Is the assumption in question 31 that the motion is streamline valid?

39   The flow of water through a water main pipe of diameter 1 m is to be modelled by water flow through a pipe of diameter 100 mm.
How should the velocities of the water in the two pipes be scaled for the flows to be dynamically similar?

40   Water is to be drained from a reservoir by a pipe of diameter 0.5 m. What is the maximum flow rate at which it is reasonable to consider the flow to be streamline?

41   Water escapes through a pipe in the base of a tank.
If the pipe has a diameter of 30 mm and the level of water in the open tank above the pipe is 4 m will the flow be streamline or turbulent?

Appendix 4.1   Hydraulic engineering

The following extract is taken from an article by M. Kendrick in the journal *Project*, September 1968 (Published by Central Office of Information).

Although in its narrowest sense hydraulics is 'the science of the conveyance of water through pipes', it is now usual to give the term a wider meaning to bring within its scope almost all aspects of the flow of water. The work which is carried out at the Hydraulics Research Station near Wallingford in Berkshire comes largely within this wider definition. This is because it is concerned with a study of the behaviour of water flowing in channels rather than pipes, its surface being open to the air and therefore subject to atmospheric pressure as, for instance, in rivers, estuaries, harbours and coastal areas.

Most of the research effort at Wallingford is directly aimed at providing practical solutions to specific engineering problems created by flowing water — problems such as the alleviation of flooding, the prevention of scour when it results in the undermining of reservoir spillways, weirs, bridge piers, the improvement and maintenance of depths in navigation channels, the feasibility of harbour development schemes, the reduction of ship movements at jetties caused by waves, and the control of beach erosion.

For the solution of these problems, each of which involves the prediction of one or more characteristics of flow, various tools are at the disposal of the engineer. His approach may be primarily by way of theoretical analysis: alternatively he may base his predictions largely on his own engineering experience gained in the field. There are, however, many problems which are not amenable to calculation or to the direct application of accepted theory, and then recourse is had to the technique of investigation by means of hydraulic models.

Contrary to popular belief, a hydraulic model is not a reproduction in miniature of a small portion of the earth's surface on which everything is correctly scaled down both in space and time. It is rather a scientific instrument which has been designed in a specific way to solve a specific problem or range of problems: indeed, a model has been likened to a calculating machine used for solving extremely difficult differential equations which describe the flow conditions in a particular area of study.

These somewhat prosaic definitions stress something that is not immediately apparent to anybody watching a model in operation, namely, that although most of the physical phenomena — flow of a river round a bend, underneath a multi-span bridge and into the sea; waves breaking on a sandy beach, carrying material along the shore and over-topping a breakwater; and so on — appear to be simulated in a strikingly realistic way, the engineer's primary concern when designing the model is to ensure that in the study area (which may be small in comparison with the whole model) those characteristics of flow which are of paramount importance will be correctly reproduced.

This can mean, although it is not a necessary consequence, that experimental results will be erroneous if based either on observations made outside the main study area or on measurements of phenomena not originally intended to be measured. This is, of course, saying no more than that having selected the type of tool for the job, the hydraulic engineer must make sure that it has been designed according to the right specifications.

What are the basic principles underlying hydraulic modelling? The primary aim is to obtain similarity between the model and its natural prototype. But similarity of what? For the former to be a perfect facsimile of the latter would require that the model system be geometrically, dynamically and kinematically similar to its counterpart in nature. This implies that between the two systems there should be equality of the ratios of all corresponding linear dimensions, equality of the ratios of all corresponding forces and equality of the ratios of velocity components at all corresponding points.

Although it is generally impossible to satisfy all of these conditions at once, almost any problem in hydraulic engineering can be resolved into the interaction of two major forces, the ratios of which can be arranged to be the same in the model as in nature.

In the simplest cases studied at the Hydraulics Research Station, the effect of gravity is much more important than the effects of viscosity, surface tension and elastic compression: similarity can therefore be achieved by making the ratio of the gravitational forces to the inertial forces the same in the model as it is in the prototype, and neglecting the less important forces.

The dimensionless quantity expressing this particular ratio, $V/(gd)^{1/2}$, is known as the Froude number, which is a function of the velocity of the current and some characteristic linear dimension such as depth. If $V_r/(g_r d_r)^{1/2} = 1$, where the subscript r indicates the model-to-prototype ratio, then the model-scales for time, current-velocity, discharge, and so on can be obtained directly from the equation.

All geometrically similar models used for studying the behaviour of hydraulic structures, such as the dissipation of energy downstream of dams, the stability of rubble-mound breakwaters under wave attack, and the design of spillways and power station intakes, are based on the Froudian criterion of similarity. Scales for this type of model can vary from 1/9 to 1/120.

5 Structure of materials

Objectives

The intention of this chapter is to introduce you to a simple atomic model of materials and to use the model to explain some of the behaviour patterns of materials. It is assumed that you can handle indices. The chapter follows from Chapters 1 and 3 of this book.

The general objectives for this chapter are that after working through it you should be able to:

(a) Describe the simple building block model for matter and the ways in which it differs for solids, liquids and gases;
(b) Explain the terms atom, molecule and mole;
(c) Calculate atomic masses and sizes;
(d) Explain how the gas laws can be obtained from a consideration of a model based on moving molecules;
(e) Explain how the stress/strain characteristics of solids can be explained in terms of their structure;
(f) Explain what is meant by the term dislocation and use this idea to explain creep, work hardening, hot working, and alloying effects.

Teaching note

Experiments appropriate to this chapter can be found in *Nuffield Physics: Pupils' Text Year 4, Nuffield Chemistry: Collected Experiments, Nuffield Advanced Physics: Teachers' Guide 1* and *Nuffield Advanced Chemistry: Teachers' Guide 1.* Further reading on crystals can be found in *Crystals and Crystal Growing* by A.H. Holden and P. Singer.

Building blocks

The photograph at the beginning of this chapter shows some examples of crystals. A particular substance such as copper sulphate, or sugar, or common salt has all its crystals of the same form. Common salt, sodium chloride, crystals always have faces such that the angle between adjoining faces is 90°. Thus some salt crystals may be small cubes, others large cubes or a number of cubes 'stuck' together. If you watch crystals growing (try watching copper sulphate growing in a few drops of solution under a microscope) you will find that they grow in a regular way (*Figure 5.1*). A cubic crystal grows by gradually expanding the cube shape.

We can offer an explanation for the form of crystals and the way they grow if we consider matter to be made up of small particles which are packed together. Spheres make convenient 'particles' to pack together to simulate crystals. *Figure 5.2* shows some of the ways we can pack spheres together in an orderly manner. The structures produced look like crystals and 'grow' like crystals.

The smell of dinner cooking in the kitchen can be detected at quite some distance from the oven. Smells seem able to move through the air. If you weigh a piece of camphor you will be able to smell it while you are weighing it and its weight will be found to be constantly decreasing. This suggests that smell is actually small particles of the substance concerned and that they somehow move through the air. It even suggests that in the camphor the particles are moving, perhaps vibrating, sufficiently for some to escape from the surface.

Liquids evaporate — particles must be moving within them sufficiently to escape from the surface. A drop of ink placed in water gradually diffuses throughout all the water: the ink particles move through the water. If you look through a microscope at very small, light pieces of matter, such as pollen or fine pieces of poster paint, suspended in

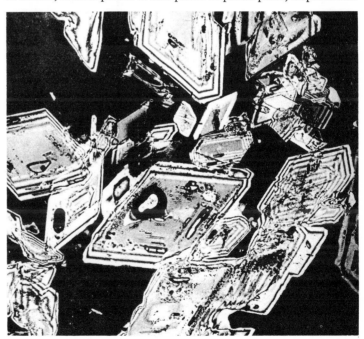

Figure 5.1 Copper sulphate crystals grown from solution on a thin glass slide

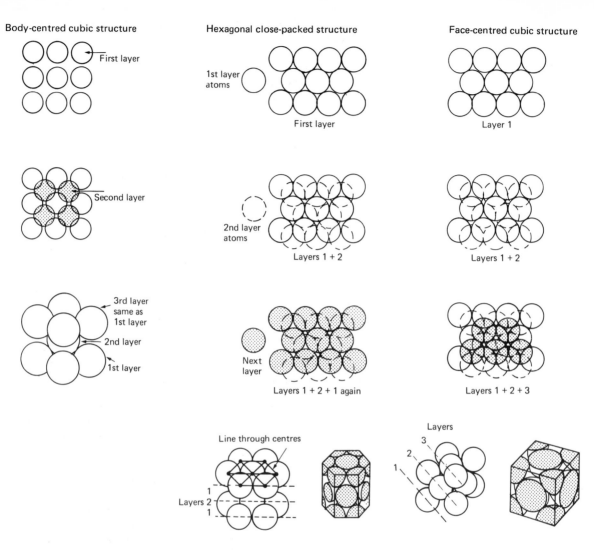

Figure 5.2

water you will see them moving. They jig around — first one way and then the other. If you look through the microscope at very fine specks of smoke in air you will see them jigging around. This effect is called **Brownian motion** after the man who first discovered it, R. Brown. This kind of movement can be explained in terms of the pieces of matter being bombarded by small particles. Sometimes there are more hitting one side of the pollen or smoke speck than the other and so the jigging around occurs.

A simple picture of matter is thus — a collection of particles. The solid consists of particles packed reasonably close together with the particles able to oscillate about their mean positions and occasionally escape. The liquid would seem to be particles less well packed and so able to move more easily. The gas with its very low density would seem to consist of particles which are able to move about very freely.

Questions   1   Explain the diffusion of ink through water on the basis of the above model.

2   Explain the movement of smells from the kitchen through the house on the basis of the above model.

Atoms and molecules

The simplest kind of particle, the basic building block, of matter which is characteristic of that matter is called an **atom**. In many instances atoms exist in clusters called **molecules.** Thus, for example, oxygen as a gas consists of molecules each of which has two atoms of oxygen. Oxygen is an element. Substances which contain only one kind of atom are called **elements**. Carbon dioxide gas has a molecule composed of two atoms of oxygen and one of carbon. Carbon dioxide is not an element — it is called a compound. **Compounds** consist of more than one kind of atom. Common salt is sodium chloride. We do not, however, talk of a sodium chloride molecule when we refer to the sodium chloride in a crystal because the crystal consists of a vast number of sodium and chlorine atoms (ions is the better term in this context, the term will be met in *Book 4: Basic Electricity and Magnetism*) all bound together in a large structure to give the block of sodium chloride we call a crystal.

In chemical reactions atoms are neither created nor destroyed but only rearranged. When carbon reacts with oxygen to produce carbon dioxide all that is happening is that the atoms of carbon and oxygen are being rearranged in the form of molecules of carbon dioxide — no atom is lost or changed. By measuring the amount of carbon that is used in a reaction with a measured amount of oxygen we can obtain the relative masses of the carbon and oxygen in carbon dioxide. This must reflect the relative masses of the carbon and oxygen atoms in carbon dioxide. If copper is heated in air it combines with oxygen to give an oxide of copper. By measuring the mass of the copper before and after the reaction the amount of oxygen that has combined with a given mass of copper can be determined. Hence the relative masses of the copper and oxygen atoms in the oxide of copper can be determined. By considering a large number of reactions between elements it is possible to draw up a table of the relative masses of atoms. Such a table is now-a-days drawn up in relation to carbon, with a particular isotope (see *Book 7: Atoms and Quanta*) of that element being given the value of precisely 12 g. Thus, for example, this gives magnesium as 24 g, oxygen as 16 g, copper as 64 g.

12 g of carbon, 24 g of magnesium, 16 g of oxygen, 64 g of copper all contain the same number of atoms. For many substances, e.g., oxygen, the molecule is the smallest amount of free substance. As the oxygen molecule has two atoms then 32 g of oxygen gas contains the same number of free particles as 16 g of atomic oxygen. The quantity of matter that contains the same number of particles as there are atoms in 12 g of carbon is called a **mole**. One mole of carbon atoms has a mass of 12 g. One mole of magnesium atoms has a mass of 24 g. One mole of oxygen molecules has a mass of 32 g.

Question   3   What is the mass in grams of one mole of water. The water molecule consists of two atoms of hydrogen with one atom of oxygen.
One mole of hydrogen atoms has a mass of 1 g.

How many particles are there in a mole? How many molecules are
there in 32 g of oxygen? How many atoms are there in 12 g of carbon?
These are really all the same question.

Lord Rayleigh dropped a small drop of oil, volume 0.9 mm^3, on
to a water surface and it spread out to cover a surface area of
550 000 mm^2. The drop of oil had a mass of 0.8 mg. The oil he used
was olive oil and one mole of this has a mass of 576 g. Thus one drop
is $0.8 \times 10^{-3}/576$ or about a 1.4×10^{-6}th fraction of a mole. The
thickness of the oil layer can be calculated from relating the total
volume of a drop to the volume of the layer when on the water surface.
Thus

$$0.9 \times 10^{-9} = 550\,000 \times 10^{-6} \times \text{thickness (m)}$$

This gives a thickness of about 1.6×10^{-9} m. If we consider this to
be the length of a molecule, i.e., the layer on the water surface was
just one molecule thick, then if the molecule was a cube (it isn't — it
is like a long stick standing upright on the water surface) the number
of molecules in the layer, and so in the drop, would be
$550\,000 \times 10^{-6}/(1.6 \times 10^{-9})^2$ or about 2.1×10^{17}. The number of
molecules in a mole is thus about $2.1 \times 10^{17}/(1.4 \times 10^{-6})$ or about
1.5×10^{23}. This is a rough, very rough, count of the number of particles in a mole. More accurate determinations of the number of particles in a mole, called the **Avogadro constant**, give a value of about
6.02×10^{23}.

Questions   4   How many molecules are there in 1 g of water?

5   What is the volume occupied by 1 g of water? Hence estimate
the volume occupied by one molecule of water.

6   Estimate the number of atoms in a piece of copper wire of
length 200 mm and diameter 1 mm. What is the size of the
copper atom if you assume that the atoms are small cubes in
contact with each other and fully occupying the space available?
The density of copper is 8900 kg/m^3.

The particle model and gases

The model of a gas has particles whizzing about in all directions in a
large amount of space. Air as a gas at about room temperature has a
density of about 1.2 kg/m^3. Liquid air has a density of about
900 kg/m^3. If we assume that the particles in liquid air are closely
packed then each particle in the gas has 900/1.2 or 750 times more
volume. If we assume that the air particles are no bigger in the gas

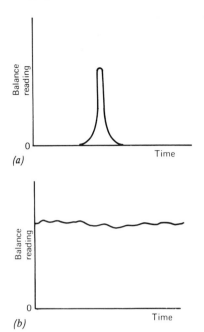

(a)

(b)

Figure 5.3 (a) One ball bearing bouncing on a scale pan, (b) a steady rain of ball bearings

than the liquid then this is extra volume in which the particles can move.

If you allow ball bearings to bounce on the pan of a balance then the balance indicates a reading. If only a few ball bearings are bouncing on the pan then the reading is just a series of impulses (*Figure 5.3(a)*). If, however, a large number of ball bearings are falling in a steady rain on to the pan then the balance reading might be almost constant, (*Figure 5.3(b)*). The bouncing ball exerts a force on the balance pan and if we have a steady rain of ball bearings an almost constant force can be exerted on the balance pan.

Suppose we have a single molecule moving backwards and forwards along the length of a box, bouncing off the end walls (*Figure 5.4*). If we halved the length of the box, i.e., halved the volume available for the 'gas', and there was no change in the speed of the molecule then we would expect the molecule to make twice as many collisions in a given time with the wall. If we had lots of molecules moving backwards and forwards along the length of the box then halving the length of the box would mean that twice as many molecular collisions would occur with an end wall in a given time. But, according to Boyle's law, halving the volume of a gas doubles the pressure. It seems reasonable to consider that the pressure is due to the collisions of molecules with the walls and thus this simple model does seem to conform with Boyle's law.

Half the volume

Figure 5.4

Questions 7 Suppose we have a molecule bouncing back and forth between the walls of a box, hitting the end walls along the normal to the wall and moving between the walls with a constant speed v.

(a) If the collision time with a wall is negligible how many collisions will the molecule make with one end wall in a time t? The distance between the walls is L.

(b) What will be the change in momentum when the molecule, mass m, hits a wall and bounces back? We have assumed that its speed is unchanged as a result of the collision and the effect of the collision is purely to reverse the direction of motion of the molecule.

(c) What is the total change in momentum in time t?

(d) So far the box has been considered to contain only one molecule, a gas will, however, contain many molecules and they will be moving in all directions. The velocities of all the molecules can be resolved into three mutually perpendicular directions and thus of the total number of molecules n we can consider there to be $n/3$ moving in the direction we stipulated for our single molecule. What is the total change in momentum in time t due all the molecules in the box?

(e) What is the force on the wall, averaged over a time t?

(f) What is the pressure on the wall if it has an area A?

(g) The volume of the box is AL. What is thus the relationship between the pressure and the volume?

(h) $pV = \frac{1}{3}nmv^2$ is the answer you should have obtained. What assumption do you have to make if this is to agree with the ideal gas equation $pV/T = $ a constant?

Questions continued

8   This follows from the previous question. The density of air at room temperature and ordinary atmospheric pressure 10^5 N/m^2 is 1.2 kg/m^3.

(a) What is the mass of gas in a volume of 1 m^3?
(b) What is the value of nm in $pV = \frac{1}{3}nmv^2$ when $V = 1$ m^3?
(c) Estimate v for air.

The above arguments lead to the idea of the temperature of a gas being related to the square of the speed of the molecules. The faster the molecules move the greater the temperature. In any gas we would expect there to be a spread of molecular speeds and thus v^2 must be taken to represent the mean of the values of the v^2 terms.

The atomic model and solids

The opening photograph to this chapter shows some crystals. You might be tempted to feel that crystals are the oddities of the solid world and that most solid materials are not crystalline. The reverse is probably the true situation — crystalline substances are very common, if by crystalline we mean any substance that has particles arranged in some orderly array. All the metals are crystalline. You may well object to this statement and say that they look far from crystalline. *Figure 5.1* shows crystals of copper sulphate growing — what would the picture have looked like if it had been taken a little later in time when the crystals had completely occupied all the solution? Would they have been like *Figure 5.5*? The surface of the galvanised iron wheelbarrow shows a thin layer of zinc on the iron surface. The zinc shows a mass of crystals, just like the result produced when all the copper sulphate crystals grow together to form a mass of crystals. Metals are termed **polycrystalline** because they are composed of a large number of small crystals rather than a single large crystal.

Figure 5.5 A galvanised iron wheelbarrow. Note the appearance of the surface. If you look carefully at such a surface you will see the crystals

The crystal patterns in the surface of many metals are not visible because of the treatment given to the surface. The patterns can be made visible by careful etching of the surface using an acid or some other reagent. The acid attacks those areas of the surface which are under local strain — the junctions between the individual crystals, or grains as they are generally called.

Figure 5.6 shows a simple 'model' of a metal structure that can be used to study the effects of pulling or compressing a metal. The model is an array of small bubbles on the surface of a soap solution. The bubbles pack together in an orderly manner and 'polycrystalline' structures can easily be produced. When the bubble raft is compressed or extended the processes that take place are: initially elastic behaviour, the raft returning to its original form when the stress is removed; and at higher stresses, the occurrence of permanent deformation. This happens when bubbles in one row slide past bubbles in the neighbouring row by an amount equal to the distance between neighbours.

If you try to make a raft of bubbles you will most likely find that the raft is not perfect, gaps occur or different size bubbles appear within an otherwise homogeneous raft or the rows do not pack together in the same way on adjacent rows (this is called a **dislocation**). Such 'faults' appear to occur with the packing of real atoms and are vital to the explanation of the behaviour of metals.

Figure 5.6 (a) The experimental arrangement, (b) a 'polycrystalline' raft of bubbles, (c) a raft with a dislocation

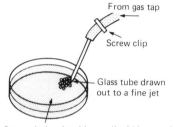

From gas tap

Screw clip

Glass tube drawn out to a fine jet

Soap solution (washing-up liquid in water)

(a)

(b)

(c)

How can we explain the behaviour of plastics? Plastics contain giant molecules, usually in long thin chains of many thousands of atoms. When polythene is pulled (*Figure 5. 7(a)*) it initially shows a small amount of elastic strain and then starts to 'neck'. In one region of the sample the width of the sample decreases to about half. The narrowed region then gradually spreads along the length until virtually all the sample is reduced in width. If the material is then pulled further it is found to have a higher modulus of elasticity. We can explain this behaviour by considering that, initially, the long molecules in the polythene are all tangled up. When the necking starts to occur the molecules are being lined-up. The polythene molecules, when the material is not being pulled, have a folded structure. When the material is pulled so that it almost breaks the molecules are considered to have straightened out — hence the higher modulus of elasticity at higher forces. It is as though we are pulling a spring. Until the spring has straightened out it shows large extensions for relatively small forces. When it is straightened out large forces are needed to produce further extensions.

A plastic like Bakelite® or Melamine® is rigid and not very easily stretched (*Figure 5. 7(b)*). In this plastic the long molecules are all tangled up but this time the molecules are cross-linked with one another. The molecules cannot untangle when a stress is applied.

Rubber consists of a tangle of large molecules (*Figure 5. 7(c)*) which changes to an orderly structure when stress is applied. This is why it is easier to pull a piece of rubber initially than when it is under a high stress. When rubber is vulcanised cross-links are made between the long molecules and so the rubber loses its elastic behaviour and becomes a rigid solid.

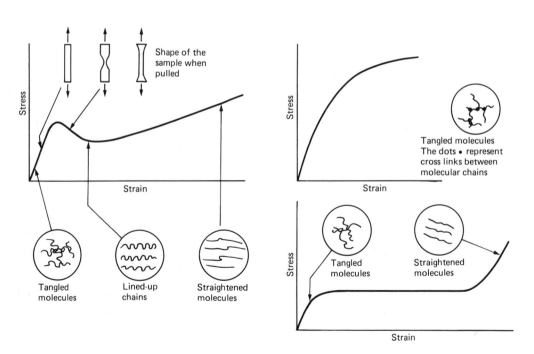

Figure 5.7 The effects of pulling (a) polythene, (b) a thermosetting plastic, e.g., Bakelite® or Melamine®, (c) rubber

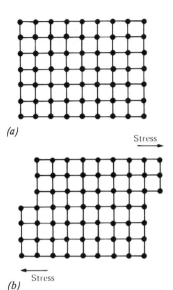

(a)

Stress →

← Stress

(b)

Figure 5.8

9   Plastics tend to suffer, more than metals, from what is called creep. If a piece of the material is under a constant force for a long period of time it gradually increases in length. Why would you consider plastics more liable to creep than metals?

10   Glass is a network of linked molecules. A piece of glass is thus essentially one large structure of linked molecules. What type of behaviour would you expect glass to have under stress?

Dislocations and the behaviour of metals

If you suspend a load by means of a strip of lead then you will find that the extension of the lead measured immediately after the load has been applied is different from the extension after, say, an hour or a day. The extension of the lead gradually increases with time. This phenomenon is called **creep**. With lead the creep is particularly notice-able. It does, however, occur with other materials.

How can we explain creep in terms of the building blocks within the polycrystalline metal? Why should the metal continue to extend when the load remains constant?

If we consider a metal crystal to be a regular packing of atoms with each atom perfectly in place then we have a crystal model something like *Figure 5.8(a)*. When stress is applied and a permanent deformation produced we have to conclude that one layer of atoms has slid over its neighbouring layer. This means that a large number of bonds between the atoms in the two layers have had to be broken and then remade, all at once.

If, however, we consider the packing of the atoms in the metal crystal to be imperfect then permanent deformation can be produced with much less stress. If you have a large carpet which is perfectly flat on the floor then it requires quite an effort to slide the entire carpet and make it move over the floor. If, however, there is a ruck in the carpet (*Figure 5.9*) then the carpet can be slid over the floor by pushing the ruck along a bit at a time. This is the type of movement that we consider takes place in the metal crystals (*Figure 5.10*). The 'ruck' in the crystal is called a **dislocation**. *Figure 5.6(c)* shows a dislocation in an array of bubbles in a bubble raft.

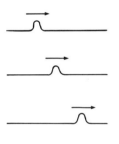

Figure 5.9 Movement of a ruck across a carpet

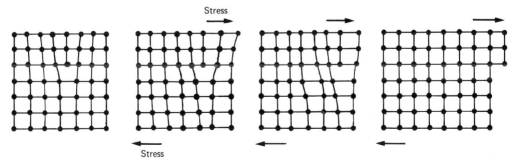

Stress

Stress

Figure 5.10 Movement of a dislocation through an atomic array under the action of stress

Increasing plastic strain

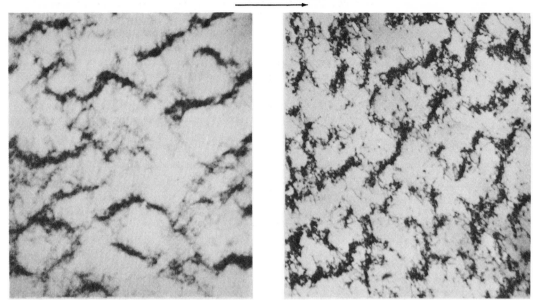

Figure 5.11 Niobium under increasing plastic strain

When stress is applied to a crystal the dislocation moves bit by bit through the crystal. Here we have an explanation of creep — the movement of one plane of atoms over another does not take place all at once. Dislocations slowly move through the crystal as gradually one plane of atoms slips into the dislocation followed later by another plane moving into the dislocation.

A polycrystalline piece of metal will contain many dislocations. *Figure 5.11* shows dislocations in niobium. The niobium has a large number of dislocations all tangled one with the other. In an undeformed metal there are typically about 10^{11} dislocations per cubic metre, in a plastically deformed metal about 10^{16} per cubic metre. Dislocations are produced as a result of plastic deformation.

What happens when two dislocations come close to one another? As will be apparent from *Figure 5.12* the atoms on one side of the dislocation are closer together and so that part of the material is in compression. On the other side of the dislocation the atoms are further

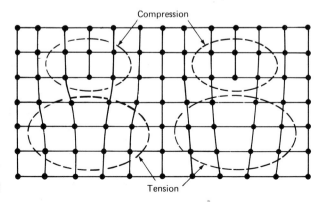

Figure 5.12

apart and so in tension. When two dislocations come close together the regions of compression can impinge on each other and hinder closer movement of the dislocations. Thus the more dislocations a metal has the more difficult it is for atomic planes to slip over each other as this involves dislocations moving through the metal and the dislocations will get in the way of each other. The more dislocations a metal has the more stress we might therefore expect to be needed to cause a given strain.

If a metal is plastically deformed a number of times, e.g., by hammering or rolling, the material becomes harder and is said to be **work hardened**. Such a material has a higher strength after the working than before. We can explain this in terms of the plastic deformation producing new dislocations and this increased density of dislocations restricting the movement of atoms in the metal.

The boundaries between the crystal arrays in a polycrystalline metal are regions of dislocations (see *Figure 5.11*) and thus the smaller the crystals in a metal the greater the density of dislocations. Thus, we might expect a metal composed of small crystalline areas to be stronger than one with large crystalline areas. Growth in crystal size can occur if a metal is heated to more than about 0.4 times its melting temperature in degrees kelvin. Thus at such temperatures the material is more easily worked. The blacksmith who formed horseshoes did so by manipulating the hot metal. There are many engineering processes that use this **hot working** of metals.

One way of hindering the movement of dislocations in a metal is to introduce 'obstacles'. **Alloying** introduces foreign atoms and distorts the normal packing of the atoms in the metal. Alloying can thus be expected to increase the strength of a metal.

Suggestions for answers

1 The ink particles are buffeted around by the movement of the water particles and are gradually knocked into all parts of the water.

2 The smell particles are buffeted around by the movement of the air particles and are gradually knocked into all parts of the house. Convection currents also play a part in spreading smell.

3 $2 \times 1 + 16 = 18$ g

4 $6.02 \times 10^{23}/18$ or about 3.3×10^{22}

5 Density = 1000 kg/m^3, hence the volume of 1 g is 10^{-6} m^3. Volume occupied by one molecule = $10^{-6}/(3.3 \times 10^{22})$ or about 3×10^{-29} m^3. This is a cube of side 3.4×10^{-10} m.

6 Mass of 1 mole = 64 g. Mass of the copper wire = $8900 \times 0.2 \times \pi \times 0.001^2/4$ or about 0.0014 kg. This is 0.022 mole. Hence the number of atoms is $6.02 \times 10^{23} \times 0.022$ or about 1.3×10^{22}. The volume occupied by an atom is $1.5 \times 10^{-7}/(1.3 \times 10^{22})$ or about 1.2×10^{-29} m^3. This is a cube of side 2.3×10^{-10} m.

7 (a) Distance travelled in time $t = vt$, hence number of collisions is $vt/2L$. The factor of 2 occurs because we are considering just one end wall and the molecule has to travel a distance of $2L$ between collisions.

(b) $2mv$

(c) $2mv \times vt/2L = mv^2 t/L$

(d) $\frac{1}{3}nmv^2 t/L$

(e) $\frac{1}{3}nmv^2/L$ as the force is the rate of change of momentum.

(f) $\frac{1}{3}nmv^2/LA = p$

(g) $pV = \frac{1}{3}nmv^2$

(h) nm the total mass of the gas has to be constant and so does v^2. This implies a relationship between v^2 and T.

8 (a) 1.2 kg
(b) 1.2 kg
(c) $v^2 = 3 \times 10^5 \times 1/1.2$, hence v is 500 m/s

9 Because the tangles can slowly unravel under stress, slowly rearrange themselves.

10 Rigid, stiff, brittle — high modulus of elasticity and a yield stress close to the breaking stress.

Further problems No suggestions for answers are given for these problems.

11 In 1665 R. Hooke stacked musket balls in piles to simulate the characteristic shapes of crystals.
In what way do such piles simulate crystals? In what way does this offer evidence for the existence of atoms?

12 A drop of olive oil of volume 16×10^{-10} m^3 put on a clean water surface spreads out to form an oil patch with an area of about 1 m^2.
What does this tell you about the size of the olive oil molecule?

13 Gold can be beaten into very thin sheets. A volume of just one-tenth of a cubic centimetre can be beaten into a sheet occupying an area 5 metres by 1 metre.
What is the thickness of the foil? What does this tell you about the size of gold atoms?

14 One mole of hydrogen has a mass of 2 g. If the density of hydrogen at 0 °C and 10^5 N/m^2 pressure is 0.09 kg/m^3, what is the volume of one mole?

15 Radium is radioactive and decays. In doing so, alpha particles are emitted. Alpha particles are helium atoms. In one year 0.043 cm^3 of helium gas was produced by 1 g of radium. From

measurements of the rate at which alpha particles were emitted it was estimated that 1 g of radium produces 11.6×10^{17} alpha particles in a year. The helium gas volume was measured at 0 °C and 10^5 N/m² pressure. Under these conditions the volume occupied by a mole of helium is 22.4×10^{-3} m². What is the number of molecules in a mole?

16 How many hydrogen molecules are there in a box which contains 10 g of hydrogen?

17 Consider a 'particles in motion' model of a gas.
How can this model explain the fact that gases exert pressure on the walls of containers?
An experiment with a pressure gauge indicates that the pressure is uniform throughout a gas in equilibrium in its container.
What does this indicate for the 'particles in motion' model?

18 How can Brownian motion be explained on the basis of the 'particles in motion' model of matter? Why is Brownian motion not normally evident with larger objects?

19 (a) A sealed container of gas has its temperature raised, the volume remaining constant.
What happens to: (i) the number of molecules in the gas, (ii) the average speed of a molecule, (iii) the pressure?
(b) Now suppose that instead of changing the temperature the pressure is changed. The volume is kept constant.
What happens to: (i) the number of molecules in the gas,
(ii) the average speed of a molecule?
(c) Now suppose that the temperature is kept constant but the volume of the container is increased.
What happens to: (i) the number of molecules in the container,
(ii) the average speed of a molecule, (iii) the pressure?

20 Nylon consists of long molecules. A feature of the production of nylon is the stretching of the material to line up the long molecules.
What type of stress/strain graph would you expect for nylon?

21 (a) Graphite is a form of carbon and is the 'lead' in the lead pencil. It is a soft material, easily transferred to other surfaces by rubbing. *Figure 5.13(a)* shows the structure of graphite. Explain how the structure leads to the properties described.
(b) Diamond is a form of carbon. It is a very hard material. *Figure 5.13(b)* shows the structure of diamond.
Explain how the structure leads to the property described.
Why is this form of carbon hard and the graphite soft?

22 Polythene in a state where it has not been subject to any stretching has a Young's modulus of about 2 GN/m². After being stretched the polythene, said to be cold drawn, has a modulus which can be as high as 70 GN/m². Explain the reason for the change.

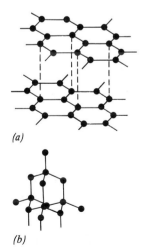

(a)

(b)

Figure 5.13 Structure of (a) graphite, (b) diamond

Acknowledgements

The publishers wish to acknowledge the following for kindly supplying illustrations and extracts and/or permission to reproduce them.

Avery Dennison Ltd — for *Figure 1.5(b)*
British Aerospace — for the illustration on page 60
British Leyland — for the illustration on page 6
British Steel Corporation — for the illustration on page 24
Camera Press — for *Figures 2.7* and *2.9*
Cement and Concrete Association — for *Figure 2.11*
J.B. Corrie and Co. Ltd. — for *Figure 5.5*
Richard Daleman Ltd — for the illustration of Evoware, *Figure 1.1*
Freeman Fox and Partners, Consulting Engineers — for the illustration on page 1
The Director, Institute of Geological Sciences — for the illustration on page 80. (Crown copyright reserved)
Professor G. Jackson — for *Figure 4.1(a)* and *(b)*
Professor A. Keller — for the original of the cover illustration. (From A. Keller and A. O'Connor (1958). *Discuss. Faraday Soc.*, **25**, 114)
Longman Group Limited — for *Figure 5.1* from *Physics 11–13* by J. Lewis and *Figure 5.11* from *The Structure and Properties of Engineering Materials* by B. Harris and A.R. Bunsell
Popperfoto — for the illustration on page 36
The Royal Society — for *Figure 5.6(b)* and *(c)*
Michael Elliott Taylor — for *Figure 1.7*

600 Magazine — for the article 'A product of elementary logic' by H. Foster
British Steel Corporation, Tubes Division — for the extracts from *SHS the Builder*
Project — for the article 'Hydraulic Engineering' by M. Kendrick, *Project* September 1968. (Central Office of Information)

Bibliography

Crystals and Crystal Growing by A.H. Holden and P. Singer, *Science Study Series No. 6*, Heinemann Educational; London (1961)
The New Science of Strong Materials by J.E. Gordon, Penguin; Harmondsworth (1968)
Nuffield Advanced Chemistry: Teachers' Guide 1 Longmans/Penguin; Harmondsworth (1970)
Nuffield Advanced Physics: Teachers' Guide 1 Longmans/Penguin; Harmondsworth (1971)
Nuffield Chemistry: Collected Experiments Longmans/Penguin; Harmondsworth (1976)
Nuffield Physics: Pupils' Text Year 3 Longmans/Penguin; Harmondsworth (1976)
Nuffield Physics: Pupils' Text Year 4 Longmans/Penguin; Harmondsworth (1978)
Project Physics Course, Handbook by F.J. Rutherford *et al.*, Holt, Rinehart and Winston; New York (1970)
Shape and Flow by A.H. Shapiro, *Science Study Series No. 20*, Heinemann Educational; London (1964)
Structures by J.E. Gordon, Penguin; Harmondsworth (1978)
The Use of Materials, Engineering Science Project Schools Council/ Macmillan; London (1975)

Index